Charles Taylor's Ecological Conversations

Charles Taylor's Ecological Conversations

Politics, Commonalities and the Natural Environment

Glen Lehman
Associate Professor, University of South Australia

First published 2015 by
PALGRAVE MACMILLAN

Palgrave Macmillan in the UK is an imprint of Macmillan Publishers Limited, registered in England, company number 785998, of Houndmills, Basingstoke, Hampshire RG21 6XS.

Palgrave Macmillan in the US is a division of St Martin's Press LLC, 175 Fifth Avenue, New York, NY 10010.

Palgrave Macmillan is the global academic imprint of the above companies and has companies and representatives throughout the world.

Palgrave® and Macmillan® are registered trademarks in the United States, the United Kingdom, Europe and other countries.

ISBN: 978–1–137–52477–5

This book is printed on paper suitable for recycling and made from fully managed and sustained forest sources. Logging, pulping and manufacturing processes are expected to conform to the environmental regulations of the country of origin.

A catalogue record for this book is available from the British Library.

Library of Congress Cataloging-in-Publication Data

Lehman, Glen, 1961–
    Charles Taylor's ecological conversations : politics, commonalities and the natural environment / Glen Lehman (associate professor, University of South Australia).
        pages cm
    Includes bibliographical references.
    ISBN 978–1–137–52477–5
    1. Environmentalism – Philosophy. 2. Environmentalism – Political aspects. 3. Taylor, Charles, 1931– – Political and social views. 4. Nature – Effect of human beings on – Philosophy. 5. Deep ecology – Philosophy. 6. Common good. 7. Democracy – Philosophy. 8. Authenticity (Philosophy) 9. Expressivism (Ethics) 10. Environmental ethics. I. Title.
GE195.L425 2015
304.201—dc23   2015005427

The book is also dedicated to the late Professor Bill Brugger and Dr. Suzanne Brugger.

Contents

Acknowledgements

Firstly, I would like to thank my parents, Douglas and Patricia, for providing a supportive environment which has enabled the production of this book. My nieces, Andrea and Renee, have also read chapters to improve the book's readability. I would also like to thank my research assistants, Andrea Brunt, Sophie Henscke and Jessie Whalen, who have spent many hours assembling Charles Taylor's bibliographic material and other tasks during the production of this book. Together with Professor Yvonne Corcoran-Nantes, Dr. Norman Porter and Dr. Michael Sullivan they supervised the project for many years.

Part I

1
Introduction

Aims and objectives

The interpretivist work of Charles Taylor is used in this book to explore the natural environment as a shared ecological and social commonality. The key focus is on the supposition that the natural world possesses intrinsic value and new political structures are needed. My preliminary thinking for this book began in response to important environmental manifestos such as *The Brundtland Report*[1] on sustainability and *The Helsham Report* in Australia about how to preserve threatened rainforests and tracts of native land.[2] One of the most contentious points concerned the economic viability of the Lemonthyme area in Southern Tasmania. The *Helsham Report*'s economic reasoning was criticised from many quarters because it submerged discussions about the political implications of the dilemma into the debate between development and conservation. In adopting an essentially economic methodology for environmental management, the report perpetuated the modernist assumption that nature stands in wait of humanity.

It soon became apparent that these environmental reports were shallow, and it became clear that a broad philosophical investigation was required, involving a range of metaphysical and theoretical questions that these reports did not address. These questions concerned humanity's impact on nature and required consideration of issues such as skirmishes over territorial waters, climate change, declining stocks of indigenous species, different rates of adaptation, despoliation of beaches and debates concerning population limits.[3]

This book endeavours to offer a different way to think about these questions by examining environmental and political concepts emanating from the eighteenth-century Enlightenment through to the current environmental thinking of Charles Taylor. Taylor has stated: 'It seems to me that every anthropocentrism pays a terrible price in impoverishment in this regard. Deep Ecologists[4] tend to concur from one point of view, theists from another. And I am driven to this position from both.'[5] In this summary statement, Taylor acknowledges that he uses ideas from Deep Ecology and theism to rethink humanity's place in the natural environment. Implicit in the statement is his quest to understand how modern approaches to environmental and social relationships have been understood in Western political theory.[6]

By way of introduction, this book on Taylor's relevance to environmental politics canvasses liberal frameworks concerning nature and objections from the perspective of Deep Ecology, far-from-equilibrium systems theory and postmodernism. It then examines those liberal frameworks in line with Taylor's counter-posing authenticity to liberal proceduralism. It goes on to explore and utilise Taylor's alternative, noting that his moral framework takes us from a purely human-centred view of ecology to a theistic social imaginary that guides *A Secular Age*. While in *Sources of the Self*, Taylor explains that he aims to reveal the environmental and social impacts of two dominant traditions of thought that shape modern political discourse.[7] The first picture suggests a conception of the subject as a disengaged agent who is 'free and rational to the extent that he has fully distinguished himself from the natural and social worlds'.[8] The second picture that Taylor develops involves deconstructing the assumptions that underpin this disengaged approach to human agency. His deconstruction of classical political thought criticises modern epistemological premises and replaces them with a picture of the human agent as actively engaged in the natural environment.[9] He explains that these critiques of epistemology often turn on this interpenetration of the scientific and the moral sphere.

I will argue that Taylor explains how his critique of epistemology develops ideas from Kant, Heidegger, Merleau-Ponty and Wittgenstein. These extremely important thinkers offer a shift away from purely disengaged pictures of the human agent. He states that there is a certain interpretivist continuity between Kant and Heidegger, Wittgenstein and Merleau-Ponty. This engagement points toward a metaphysical analysis to further challenge modern political disengaged approaches to human agency.[10] It is then a relatively simple extension to explain how our conception of human agency impinges adversely on the natural environment. That is, our political systems become infatuated with economic growth at the expense of creating improved relationships with other communities, nation-states and other entities living in the natural environment. My argument is that interpretivism has the potential to inform new community structures and to change thinking thereby moderating our adverse impacts on the natural environment.[11] Moreover, in the chapters that follow I examine the debate between interpretivists and critical theorists who have made important inroads into environmental debates.[12] The important arguments that they make involve the need to recognise and understand that interconnections between our understanding of commonalities, difference, location and place.

However, before examining these challenges to the modern picture of disengagement involves an openness to the world through an explanation of the intricate connections between people and the natural environment. This leads Taylor to his emphasis on how people's identities are formed within cultural, environmental and linguistic traditions that frame the factors which have significance in their lives. Clearly, one crucially significant factor is the way people relate to the natural environment. Taylor's thought, therefore, is relevant to considerations of what it means to be a person as if the processes of the natural environment are not important.[13] In particular, Taylor's work on authenticity, interpretation and secularism maintains that the most promising ecological argument must engage citizens so that they are involved in community environmental decisions.

These issues are outlined in two principal sections of the book. Part I comprises five chapters examining the connections between Taylor's work on agency, interpretivism, metaphysics and social imaginaries in environmental contexts. Part II has seven chapters of which three chapters are on Taylor's environmental and political engagement with critical theory, liberalism and utilitarianism and four chapters examine a number of critical perspectives on Taylor's work emanating from various postmodern, radical and social approaches to environmental politics.

2
Basic Issues in Taylor's Philosophy

This chapter outlines some key issues and definitions in Taylor's philosophy to set the scene to understand our place in the natural environment. The focus of this chapter is on interpretivism, the public sphere, and the processes of secularism.

2.1 Environmental conjectures and refutations: preliminary issues

In *On Walden Pond*, the famous environmentalist David Henry Thoreau claimed that 'in wildness is the preservation of the world'.[1] This famous maxim challenges the modern supposition that answers to humanity's environmental dilemmas can be found through human ingenuity and technology. Taylor has often stated that he agrees with Thoreau, and thus he has explored how the relationship between humanity's social and natural environments has been understood in Western political philosophy since the Enlightenment.[2]

Taylor offers an interesting approach that can be applied to environmental politics by using modern debates concerning how we interact in the natural environment to challenge how humanity thinks about relationships with the natural environment. In this regard, Taylor uses the interpretivist ideas he has developed to critique political perspectives that are based on instrumental ways to reason about the natural environment, because they invariably sever the connections between the social and natural environment. This is why many environmentalists concur with Deep Ecologists and other social environmentalists concur with theism. He combines ideas from both traditions, which leads to an openness to different values that may emanate from within us or from the whole of nature.[3]

The environmental argument developed in these chapters draws upon the interpretivist framework employed by Taylor to explore the Enlightenment's conception of rationality and reason, applying it to political matters central to environmentalism. Furthermore, I use interpretivism to explore the Enlightenment's commitment to autonomy and the way it evaluates the natural environment as a good to be conquered and mastered.[4] I examine the Enlightenment's stress on autonomy and the way it submerges political and ecological values. My environmental argument affirms what Taylor calls *authenticity*[5] – extending our understanding of autonomy to consider processes of human self-realisation through exploring humanity's 'being-in-the-world'.[6] Autonomy refers to human freedom and the ability to be impartial, but authenticity demands contextualisation of that autonomy. The notion of authenticity helps in considering how Aristotle's 'practical reason' can illuminate values other than those which reflect instrumental considerations.[7]

In particular, Taylor's work on authenticity informs this book because I believe that the most convincing ecological argument must be built on interpretivist premises. By way of preliminary clearing, my argument canvasses liberal frameworks concerning nature and objections to them from the perspective of Deep Ecology, far-from-equilibrium systems theory and postmodernism. The environmental perspective examines modern liberal frameworks counter-posing Taylor's authenticity to the liberal infatuation with proceduralism. I go on to explain Taylor's alternative, noting that Taylor's moral framework takes a theistic direction. In this chapter I ask: can Taylor's approach be extended to escape anthropocentrism? I explore whether we fully understand our impact on nature. My key argument revolves around Taylor's attempt to preserve threatened values by understanding perspectives on authenticity and expressivism which provide a radical reworking of humanity's understanding of being-in-the-world, and a starting point for rethinking the way individuals and communities ought to be dealing politically with ecological crises.

This chapter outlines key structural issues in Taylor's political philosophy in four key sections following this introduction to the chapter. Section 2.2 introduces Taylor's interpretivist framework as it applies to his environmental and political work. Section 2.3 extends Taylor's interpretivism to search for meaning and intrinsic value in the natural environment. Section 2.4 introduces Taylor's ideas on exclusive humanism, instrumentalism, and secularism in an environmental context. Section 2.5 concludes the chapter by linking Taylor's work on secularism to environmental politics in the public sphere: these arguments involve debates with the communicative theorist, Jürgen Habermas.[8]

2.2   The relevance of Taylor's interpretivism to environmental politics: a succinct consideration

I will use Taylor's work on liberalism and interpretivist frameworks to construct a bridge between democratic, ethical, and ecological perspectives. This leads to a fusion between a broad liberalism and interpretivist ideas, which involves developing common ground to recognise cultural and environmental issues. This fusion of perspectives in the search for a common ground is done in a spirit that aims to moderate the dominant anthropocentric attitude toward the natural environment. More broadly, the ecological perspective developed here owes much to Taylor's interpretation of Martin Heidegger's argument that people are thrown into a world saturated with meaning and value. It owes a debt to Merleau-Ponty's explorations into how we perceive the world in which we live.[9] People's identities are formed within cultural and linguistic traditions that frame the factors which have significance in their lives. Clearly one crucially significant factor is the way people relate to nature.[10] He argues that Heidegger's work on language is relevant to considerations of what it means to be a person in the modern world.[11]

Furthermore, it is important to remember that Taylor has been influential in developing the art of interpretation and applying it to issues associated with environmental politics, multicultural recognition, and international relations.[12] He has explored the problems that arise when we solely rely on modern theories of knowledge that emphasise procedural political solutions by using the interpretivist framework. Moreover, there is a critical framework within the interpretivist perspective, a fact that has often been overlooked by critics who say that Taylor's theological commitments leave him whistling in the dark.[13] The interpretivist argument is that modern environmental and political thinking has narrowed how the natural environment is conceptualised. Interpretivism also involves examining current theories of knowledge and explaining how they have shaped the societies in which people live and their relationships with the natural environment. This interpretivism examines the limitations and possibilities within the dominant theories of knowledge as they apply to environmental and social dilemmas. Central to Taylor's interpretivist framework is an argument that modern economic and political sciences have created environmental and social practices that narrow how the environmental and social dilemmas of modern societies are perceived.

Therefore, Taylor's interpretivist framework transcends modernity's infatuation with procedure and escapes the visions Max Weber conjured with the 'iron cage' of bureaucracy.[14] On this view, interpretivism moves beyond current political methods that are outgrowths from economic and procedural strategies. These arguments reflect Taylor's dissatisfaction with approaches associated with instrumental reason and procedural liberalism and invite rethinking what would be a full life and how people could strive towards it. To save the natural environment and recognise values of political significance, Taylor gives prominence to the power of a dialogic society. He is particularly critical of political and procedural approaches that are insensitive to significant cultural and environmental values. Taylor has recently pointed out that his key works, *Sources of the Self* and *A Secular Age*, were inspired by a quest to better understand the plurality of world-views that have been used to explain the meaning and intrinsic value in the natural environment.[15] Moreover, *Sources of the Self* was underpinned by the belief that the natural environment and humanity are entwined. When it comes to articulating different visions of the natural environment, these critical issues require careful attention in a political sense. Furthermore, Hittinger was one of the early reviewers to observe that the Deep ecological and theistic dimensions in *Sources of the Self*[16] left him reliant on an external and transcendental God.[17] That is, 'there is a theology that is crucial to, and yet left inarticulate in, Taylor's treatment of these issues'.[18] These critical responses to Taylor, however, do not engage with his philosophical and speculative quest to reveal humanity's relationships with the natural environment.[19] One of Taylor's principal aims is to reveal the anthropocentric limitations within exclusive humanism, which cuts people off from the possibility that there exist other sources of the self. Therefore, the interpretivist and narrative dimensions in his work are designed to open us to the splendour and value within the natural environment. That is, the natural environment cannot be reduced to physical causes and effects but contains meaning and value that language-bearing agents appreciate and recognise. Taylor argues that the meaning and intrinsic value in the natural environment is submerged by those instrumental approaches that deny the existence of values independent of a valuing subject.

2.3   Extending Taylor's interpretivism to environmental politics: the search for meaning and intrinsic value

Drawing on Hegelian and Romantic traditions, Taylor not only challenges the idea that nature is there to be consumed, but doubts whether humanity is capable of mastering the natural environment, which is far more intractable than most liberal assumptions allow.[20] Far from mastering it, we may perhaps never fully understand it but at least Taylor's proposals for informed discussion might highlight what we do know about nature's processes. This is evident in his earliest work where he explores limits to growth and the possible shape of steady-state communities.[21] Here he is at one with ecologists, deep and shallow, though he tends to concentrate more on developing ways to think about how to organise modern communities so that citizens are aware of the decisions being made on their behalf.

Any reader of Prigogine and Stengers' *Order Out of Chaos*[22] and Stengers' *Cosmopolitics*[23] might detect a contradiction between nature's intractability and the advocacy of steady-state communities. They maintain that systems are rarely in equilibrium and that far-from-equilibrium situations are the normal state of affairs:

> We know now that societies are immensely complex systems involving a potentially enormous number of bifurcations exemplified by the variety of cultures that have evolved in the relatively short span of human history. We know now that such systems are highly sensitive to fluctuations. This leads both to hope and a threat: hope, since even small fluctuations may grow and change the overall structure. As a result, individual activity is not doomed to insignificance. On the other hand, this is also a threat, since in our universe the security of stable, permanent rules seems gone forever.[24]

Is it possible here to make use of the separation between ontology and advocacy?[25] Surely, new scientific methodologies, such as those pioneered by Prigogine and Stengers, offer an opportunity to extend interpretivism in a new interpretivist and post-liberal direction. They define 'dissipative structures' as non-equilibrium spatial structures which occur beyond the point of bifurcation and offer a new set of phenomena whereby 'we may have oscillating chemical reactions, non-equilibrium spatial structures, or chemical waves'.[26]

From interpretivist quarters, the issue is to recognise how ecosystems are 'dissipative structures maintained by the matter and energy flowing through them'. Once formed, these structures, in order to keep their shape, need to dissipate entropy so that it will not build up within the system and destroy it (or return it to equilibrium). Because dissipative structures produce high levels of entropy, they require high inputs of matter and energy.[27] The structures, moreover, are prone to random fluctuations which might result in what is known as a 'bifurcation point', where the structure reorganises itself at a higher or lower level of complexity (a possibility which cannot be predicted).

Here Taylor's interpretivism must recognise that a dissipative structure emerges from a chaotic situation of change, where many options exist but only one path can dominate. But, because of the nature of the disturbance theorists cannot predict what will occur. Prigogine and Stengers' argue that the classical understanding of science is no longer relevant because systems are never close to equilibrium – they are dissipative structures. Accordingly, physics and biology respond to the environment, take in energy and grow. Classical thermodynamics has little to say about this form of negative entropy which has implications for the construction of human communities. This understanding of thermodynamics, however, attempts to unite both the second law of thermodynamics (the tendency toward entropy) and the development of order out of chaos. The implication of this argument is that the mechanistic approach of the Industrial Revolution was obsessed with equilibrium or near-to-equilibrium situations. 'The point is that these systems tended to break down, just as the archetypal steam engine wore out. One saw entropy everywhere. If, however, one looked at far-from-equilibrium situations, one could see order generated everywhere.'[28]

The natural environment is but one of many systems that are like a dissipative structure. Moreover, the dialectical and political implication of this new approach to science involves healing the rift between scientific and philosophical interpretations of nature's value. In the light of these extensions to our understanding of social equilibria it is necessary to abandon the totalisation aspired to by Hegel. But it will be recalled that Taylor retrieves Hegel's historical method to consider the relationships between particular phenomena and the political structures which articulate common purposes. Taylor, like Prigogine and Stengers, recognises these problems in Hegel's system and avoids the totalising dilemma through an extension into our understanding of democratic structures. He accommodates political and environmental differences through a political structure which works toward the ideal of authenticity, which is not a fixed end-point but involves a normative approach that provides a critical reflection on the relationships between people and nature.[29]

Interestingly, Isabella Stengers' recent work on *Cosmopolitics* also helps us in regaining a glimpse of the 'scientific' relationships between the social and the ecological spheres together with the possibility of reconciling humanity's relationships with the natural environment.[30] She shares Taylor's Hegelian terminology that it is through practical reason in broader public spheres that we are able to avoid the type of criticisms levelled at 'shallow green' ecology, and anthropocentrism which submerges awareness of humanity's place in nature.[31] Realising the convergence of his views with part of the Deep Ecology agenda Taylor points us toward his work on environmental and cosmological imaginaries which I examine in Chapter 4.[32]

Yet, while concurring that nature possesses intrinsic value, reflected in the sense of awe that we feel when confronting it, he distances himself from Deep ecologists who define value independent of human moral categories. Taylor emphasises the role of dialogue, language and perception involved in our interactions with the natural environment. An interpretivist perspective, therefore, maintains that it is important to consider the dialogic processes through which citizens are provided with opportunities to learn about the good society through democratic discussion. We shall see in the chapters that follow that meaning and intrinsic value are important concepts that must be given consideration in modern public spheres.[33]

Speculating, Taylor would probably concur with Aldo Leopold's famous environment perspective that humanity should learn to live from the land, but would add a role for public spheres to articulate ecological concerns.[34] He certainly avoids the tendency, common among Deep ecologists, to rank ecosystem interests higher than those of humanity. Taylor's interpretivism, in common with the environmental theorist Bryan Norton's criticism of Deep Ecology, maintains that nature has value for itself, rather than in itself.[35] Norton argues that environmentalists searching for intrinsic-based values, such as J. B. Callicott, cannot locate the intrinsic value in nature independently of present language systems and structures.[36] Norton's solution is a pragmatic ecological position and uses John Rawls' idea of an overlapping consensus.[37] Enough has been said above to see how Norton's ideas could be bent in a Taylorian direction to link the ecological and the social. Taylor, however, a staunch critic of postmodernism, would be wary of the implication of Norton's advocacy of a new 'postmodern' world-view where people respect and operate within nature.[38] Perhaps, though, Norton's 'pragmatism' is similar to the postmodern position but it is not one of the extreme variety where diversity and difference are allowed to trump all conceptions of the good. Extending interpretivism to ecological politics leads simply to the supposition that the community is related to nature, where nature is not something foreign, strange and alien.[39]

2.4 A *Secular Age*: the rise of exclusive humanism

In recent work, such as *A Secular Age*, Taylor offers a master narrative using literary and philosophical visions to explain the forces and social factors that have led to current environmental and social predicaments. Taylor has often stated that he does not offer universal solutions, but rather a speculative thesis about individual well-being. In Taylor's early work, for example, he argues that Heidegger's work offers a means to explore the value in the natural environment, which is discussed later in Chapters 7 and 8. In recent work he examines various approaches to the issue of secularism which can be used to understand our environmental concerns. He focuses on what he calls 'closed world' secularism that leads to instrumental and procedural thinking, which maintains that only human values count as valid knowledge.

Of course, in his *Secular Age*, he does not argue for a dismissal of secularism altogether. Indeed, his art of interpretation or political interpretivism is relevant to environmental debates in our secular age in the manner it explores how communities ought to be dealing with the environmentally significant issues confronting them. Interpretivists are critical of Western thinking whose ability to solve cultural and environmental issues is limited by procedural and technical methods. Taylor's interpretations examine how these environmental and political traditions have created current political and social structures. He explains:

> Not all critics attacked on all axes, of course, but what they had in common was the sense that the danger which awaits us in our culture takes a certain form. We are tempted to draw the limits of our life too narrowly, to be concerned exclusively with a narrow range of internally-generated goals. In doing this we are closing ourselves to other, greater goals. These might be seen as originating outside of us, from God, or from the whole of nature, or from humanity; or they might be seen as goals which arise indeed within, but which push us to greatness, heroism, dedication, devotion to our fellow human beings, and which are now being suppressed and denied.[40]

Here Taylor's work on the processes of secularism and the role of theism can be interpreted as offering two different, but complementary, political frameworks to understand the contours of modernity. In *A Secular Age* he now offers an exhaustive account of the rise of a secular culture and the costs it can create for our being-in-the-world. From the narrative structures in that work his aim is to explore our being and place in the natural environment.

Furthermore, the dominant Western-inspired thinking assumes that the scientific method is the optimal means to order human interactions with the natural environment. Accordingly, the art of interpretation offers a critical perspective to examine current social structures and their effects on the natural environment. Many approaches have been used to examine human agency, on which modern European economic and political civilisation has been built. Taylor's interpretations examine how these traditions have created current political and social structures. His recent work focuses on how a certain procedural construal of how we acquire knowledge has gained sway. The relevance to environmental politics, I contend, focuses on what can happen when our culture takes a certain economic path that devalues anything other than pricing and trading mechanisms. As Taylor notes the limits of our life are then drawn too narrowly and we become focused on a narrow range of internally generated human goals. These are the goods of accumulation and transformation, which assume that humanity has usurped the need for Platonic forms, the natural environment or gods.

Therefore, from the direction of Continental philosophy he uses the work of Merleau-Ponty and Heidegger to argue that people are directly engaged with the natural environment and not separate from it. One implication from this way of thinking is that our critical ethical and governance structures are rendered ineffective in the moral malaise of modernity. A collective amnesia grips society where the role of common and shared values are forgotten; hence, good practice needs to be nurtured and good reasons need to be provided. Taylor returns to religion to better understand the notion of a social imaginary. He explains:

[A] pattern of practices that gives a certain shape to our social imaginary. Religion – or, as Durkheim liked to put it, the senses of the sacred – is the way we experience or belong to the larger social whole. Explicit religious doctrines offer an understanding of our place in the universe and among other human beings, because they reflect what it is like to live in this place. Religion, for Durkheim, was the very basis of society. Only by studying how society hangs together, and the changing modes of its cohesion in history, will we discover the dynamic of secularization.[41]

In *Varieties of Religion Today*, this theme is developed to explore the dominant philosophical trends in modernity and offers an implicit critique of the assumption that the market is free. He uses hermeneutic-inspired interpretivist reflections on secular processes to explain the retreat of common values that bring people together despite their differences. It is important therefore to remember that Taylor finds commonalities in various religious practices that have been submerged by processes of secularisation. Ultimately these processes reduce the public sphere to market relations.

For Taylor, neither of these interpretations of modernity nor their secular processes are attractive. This is because modern secularism seems unable to acknowledge that there is a range of religious and spiritual beliefs that empower people's lives which put us in touch with the natural environment. These beliefs are the internal goods associated with virtuous practices that influence what is worthwhile while potentially empowering social institutions to pursue sustainable outcomes. Taylor does not suggest that there has been a straightforward decline in religious belief and thus depicts secularity in a different way. He states:

> Secularity closes old, and opens new, avenues of faith. What it doesn't seem to allow, however, is a new 'age of faith', that is a time of universal belief, where the undeniable common experiences that no-one can escape bespeak some commonly agreed spiritual reality. That seems no longer possible in a secular age.[42]

In this passage, Taylor's political view emphasises why common goods are needed for well-functioning and secular societies.[43] The interpretivist argument is that nature is one such good, but one that the dominant scientific revolution has considered to be devoid of intrinsic value.[44] This is because the dominant scientific view relies on objective, disengaged, disembodied knowledge to explain our place in the natural environment. Taylor argues that this dominant scientific view leads to closed world systems. That is, the prominence of the scientific method has created a culture that conceptualises the natural environment as a good to be controlled and dominated. The purely technical scientific approach leads to a depiction of nature as a realm devoid of meaning and perpetuates an instrumental social imaginary.

For Taylor, the critical analysis of a secular age leads to a focus on the common purposes that exist between all entities. The extreme philosophical position that Taylor attacks involves an assumption that there is a strict boundary between people and the natural environment. This assumption denies the possibility that the natural environment contains intrinsic value that can influence people's life plans and the opportunities available to them. Interpretivism, therefore, is defined by what it aims to negate in the domain of modern epistemology. In the domain of epistemology there is a quest to explain that there are no ultimate foundations, nor are there any necessary explanatory factors that place the mind before the natural environment. In the formulation of knowledge there is the supposition that our skills of perception put people directly in the natural environment. The interpretivist approach, in this regard, emphasises the role of discourse and language in reformulating how communities interact with the natural environment.

2.5   Taylor's work on enchanting communicative spheres: informing relevant publics

Central to Taylor's environmental visions and moral theory is his important work on the public sphere. He emphasises that the role of the public sphere is to create awareness of values other than those determined by market forces. One way to introduce the idea of a public sphere is to return to Taylor's review of Habermas's *The Structural Transformation of the Public Sphere*.[45] Taylor challenges Habermas's theory of community, language and the public sphere to illustrate how subtle accounts of secularism (concerned with the secularisation of public spaces, the decline in belief and practice) are superseded by newer versions of secularism that allow space for belief and spiritual accounts.

Taylor battles against those who want to return to a pre-humanist framework of the good life, and defends aspects of the Enlightenment against reactionary closed world systems and forces. His interpretivism, through dialogue and dialogic structures, combines and retains the features of these political systems that are worth saving. In this regard, Taylor questions Habermas's reliance on a procedural approach to create reconciliation and recognition with other interlocutors concerning matters of significance such as the natural environment. He explains that in eighteenth-century Western Europe a new concept of public opinion emerged where commonalities were developed. Taylor points out that coming to grips with the idea of a common space is difficult. Moreover, the concept of a common space cannot be expressed using empirical and scientific methodologies. Taylor continues that the common space must be given full consideration in our collective deliberations because it is associated with factors that shape community and identity.[46] Taylor explains:

> What is this common space? It's a rather strange thing, when one comes to think of it. The people involved here have by hypothesis never met. But they are seen as linked in a common space of

discussion through media – in the 18th century, print media. Books, pamphlets, newspapers circulated among the educated public, vehiculing theses, analyses, arguments, counter-arguments, referring to and refuting each other. These were widely read, and often discussed in face-to-face gatherings, in drawing rooms, coffee houses, salons, and/or in more (authoritatively) 'public' places, like Parliament. The sensed general view which resulted from all this, if any, counted as 'public opinion' in this new sense.[47]

The people involved in the public sphere have never met but are linked in a common space of discussion which is opened up by new media mechanisms. Interestingly, this is something both Habermas and Taylor are exploring on their *Immanent Frame* discussion site. The *Immanent Frame* moderators are well aware of the danger that discussion pages may degenerate into never-ending disputes.[48] It is probably for such reasons that all viewpoints entered on that public site are subject to principles of reasonable discussion.

A further caveat is in order, however. Both Habermas and Taylor have defined the public sphere in different ways. Nevertheless, both agree that the public sphere began to expand in the eighteenth century when print media and new forms of communication emerged. It will be recalled that for Taylor the general view which resulted from this process was deemed 'public opinion'. This was first expressed by Charles Fox to the British House of Commons:

> It is certainly right and prudent to consult the public opinion ... If the public opinion did not happen to square with mine; if, after pointing out to them the danger, they did not see it in the same light with me, or if they conceived that another remedy was preferable to mine, I should consider it as my due to my King, due to my Country, due to my honour to retire, that they might pursue the plan which they thought better, by a fit instrument, that is by a man who thought with them ... but one thing is most clear, that I ought to give the public the means of forming an opinion.[49]

Here the legislative forum informs public opinion and exposes it to the pressure of public discourse, thereby illustrating the notion that legislation should ultimately bow to public opinion. Now this space provides an opportunity not simply to shape legislation but also to become involved in the common public sphere, thereby contributing to the common values that make up a society.

Yet, in recent times there has been a decline in developing the common goods in the public sphere, in spite of the fact that even back in the eighteenth-century print media opened up this new space and offered a hope that new ways to inform communities were imminent. In this regard, Taylor integrates ideas from different traditions in a fusion of horizons offering a broader interpretivist approach to environmental and social politics. The approach aims to create shared commonalities which open us to a space of reasoning that includes the intrinsic value of the natural environment. This common space, opened up by our ability to express thoughts in language, provides different ways to consider how to visualise the natural environment and appreciate our place in it. More fundamentally, a procedural political system has the potential to create a bureaucratic structure that imprisons people within an iron cage from which they cannot escape. Taylor responds to Habermas and suggests a return to an earlier way of moderating power in a public forum – the *ekklesia* – which would be called for in the Greek city states when their governments had become too corrupt and oppressive.[50] This was an assembly outside the civil authority of the city, and if enough people came out and refused to accept the existing centralised civil authority that government would collapse. Non-participation has been a successful and peaceful means to free people from oppressive civil authority throughout history. This is because it is ultimately people who have the means to expose and remove governing bodies that stifle the human condition and our ability to appreciate the natural environment.[51]

In these matters, Taylor challenges critical and radical theory to create a political thinking that accommodates complexity and examines dilemmas posed by the forces and powers of corporatisation, harmonisation and globalisation.[52] These forces narrow the role of the public sphere. This is not just an argument for more sites to discuss and narrate the failures of capitalism, but an explicit call to understand the social forces and powers that have resulted in the malaise of critical and radical political strategies to create environmental and social change. Interpretivism points towards the development of civil societies where patterns of meaning in communities reflect structures and political pathways. That is, only when it is appreciated that these forces and powers create our current predicament can we hope to find ways to combine the divergent goods on offer. The politics of change have become marginalised in the quest to facilitate free markets in globalising spatial structures.

3
Taylor's Interpretivism, Knowledge and the Natural Environment

This chapter explores the dominant Western beliefs about humanity's relationship with the environment, and how current environmental issues have been shaped in procedural and representational ways to conceive the natural environment.

3.1 Introduction

Taylor has explained that his interpretivist framework was inspired by counter-culturists such as Johann Gottfried Herder[1] and Johann Georg Hamann[2] and modern environmentalists such as David Henry Thoreau[3] and Aldo Leopold.[4] These theorists challenged key elements of the Enlightenment and emphasised the value of modern wildness and spirituality.[5] In Taylor's work on the reconciliation, these goods are important to our identity which can be extended to environmental issues. Taylor has long recognised the complexities involved in these debates and develops links between our powers of perception and language as a means to explore our place in the natural environment.

Taylor's work on interpretivism, foundationalism and the politics of recognition is based on a fundamentally different means to engage the natural environment. The global issues of significance confronting modern communities involve factors such as global warming, climate change, and wasteful and duplicitous consumption, to name a few. Political interpretation or interpretivism, therefore, is about exploring the current responses to these issues and examining how to reform existing political institutions in a globalising world order. In utilising interpretivism it is possible to examine the way our societies have come to deal with environmental and social issues.

This breadth and diversity in Taylor's work reflects a complex political strategy that aims to overcome procedural political approaches that dominate today's political methodologies. Of course, complexity itself can lead to contradictions and dilemmas within Taylor's massive body of work, but that criticism overlooks how practical reasoning is used to work through the differences by using informed, reasoned and tolerant debate. In this sense, interpretivism explores and identifies the common purposes that exist between people and the natural environment. The critical values needed for people's lives cannot necessarily be articulated and expressed using procedural scientific and technical methods. This search to understand the value in the natural environment owes much to Taylor's discussion of interpretivism and his attempt to overcome the alienation of modern life.[6] Taylor's interpretations relate considerations of broader environmental issues associated with the quality of life to general cultural and social relationships.[7]

Central to interpretivism is a political approach that focuses on commonalities and what is a good society. The aim is to recognise significant issues associated with people's strong evaluations such as the importance of nature and wildness. These basic interpretivist issues are outlined in three sections following this introduction. Section 3.2 defines interpretivism and relates it to environmentalism. Section 3.3 examines Taylor's work on engaged agency to break the division between how we think about the natural environment. Section 3.4 concludes the chapter with some environmental dilemmas that emerge from this discussion.

3.2   Defining interpretivism and Taylor's evaluative framework

It is worth remembering that interpretivism is part of Taylor's hermeneutic approach. Hermeneutics was initially used to examine religious texts. In recent times, this approach has been extended by thinkers such as Gadamer, Heidegger, Sandel and Taylor to examine issues confronting modern communities.[8] From these authors, Taylor develops the view that dialogue, discourse and the public sphere are critical to the development of good structures of governance. However it is important to define interpretivism in terms of Taylor's work on interpretation in the social sciences. Taylor explains:

A successful interpretation is one which makes clear the meaning originally present in a confused, fragmentary, cloudy form. But how does one know that this interpretation is correct? Presumably because it makes sense of the original text: what is strange, mystifying, puzzling, contradictory is no longer so, is accounted for. The interpretation appeals throughout to our understanding of the 'language' of expression, which understanding allows us to see that this expression is puzzling, that it is in contradiction to that other, etc., and that these difficulties are cleared up when the meaning is expressed in a new way.[9]

Taylor uses the term interpretivism as representative of a particular approach critical of contemporary liberalism, positivism and proceduralism and how they emerged from a disengaged picture of our relationships in the social and natural environment. Here I argue that the interpretation performs the same function as 'the art of interpretation' and 'the political interpretation'. I prefer the term interpretivism as it reflects Taylor's view that language reflects our humanity as an ongoing conversation. My understanding of interpretivism emerges in our interaction and utilisation of the concept itself. That is, self-interpretations and interpretation cannot 'be described or validated in a way that makes them immune to further questioning or subsequent reinterpretations'.[10]

Taylor's work on interpretivism is part of his evaluative framework which searches for meaning and intrinsic value in the natural environment. From an interpretivist perspective, a series of arguments have been advanced that Western political theory has created a political position that humanity can control and manage the natural environment. The application of a procedural and technical method to the humanities can submerge an appreciation of the moral frameworks more applicable to human sciences. For Taylor, interpretivism 'in the sense relevant to hermeneutics, is an attempt to make clear, to make sense of an object of study'.[11]

The aim of interpretivism in environmental contexts is to make clear our seemingly contradictory attitudes toward the natural environment. In part, Taylor's interpretivism has engaged with market-inspired strategies that are unlikely to reflect the contingencies of particular locations and significant values when constructing environmental policies. Therefore, this interpretivist argument develops a line of thinking that differs from market-based and neoconservative approaches that are dominating modern political debates over the utilisation of the natural environment. For interpretivists, such as Taylor, it is our moral frameworks that provide the means through which people perceive and understand the world. This gives rise to a set of choices for humanity one which involves the dilemmas confronting humanity which involve expressive unity with the natural environment and to radical individual autonomy. While we may not be able to fully combine individual and environmental values the Romantic rebellion continues undiminished. Taylor refers us to what he calls the first great synthesis which was meant to resolve our central environmental dilemmas. Central to Taylor's interpretivism is the romantic notion that nature possesses intrinsic value which is worth preserving. He notes that the romantic rebellion continually recurs:

> For the two powerful aspirations – to expressive unity and to radical autonomy – have remained central to preoccupations of modern man; and hope to combine them cannot but recur in one form or another, be it in Marxism or integral to anarchism, technological Utopianism or the return to nature. The Romantic rebellion continues undiminished, returning ever in unpredictable new forms – Dadaism, Surrealism, the yearning of the 'hippy', the contemporary cult of unrepressed consciousness. With all this surrounding us we cannot avoid being referred back to the first great synthesis which was meant to resolve our central dilemma: which failed but which remains somehow unsurpassed.[12]

To affirm morality's voice within us and to sympathise with romanticism, however, is not the same as affirming nature's intrinsic value independent of human beings. This would be a step Taylor is unwilling to take because his theory of language defines value through communication, debate and dialogue. Of importance to his position is the notion of 'insight' into the values of different communities which might bring about awareness that the natural environment is a resource which affects all members of communities and their society.[13]

On Taylor's view, humanity's power of reason and rationality provide the human race with objectifying powers that transform the natural environment. This choice that confronts humanity involves humanity's power to radically transform the planet for economic purposes, or to synthesise our lives in accordance with the ecosystems in which we live. This argument does not mean that we simply follow nature's spontaneous processes as some Deep ecologists and Earth firsters might advocate. It does imply, however, that people have the ability to choose different life plans and that these choices are ultimately for communities to make.

Within Taylor's interpretivism is the supposition that ethics and moral theory must be extended to explore not only differences but also what is environmentally significant. The aim of interpretivism is to reinvigorate moral sensibility between people and then recognise our embeddedness in the world. In responding to some recent critics he explained that this was one of the dilemmas that motivated his *Sources of the Self.* The interpretivist dimensions associated with the provision of practical reason offer scope to design new and different means to relate with the natural environment. There is, then, no indubitable reason why our communities should be limited by a path dictated solely by instrumental reason and unrestrained consumerism. It is probably for reasons similar to these that Martha Nussbaum suggests that we rethink our ethical place in the world and offers the useful idea that humanity has the means to act as the guardian of the earth.[14]

However, before examining key interpretivist ideals it is useful to remember that they explore not only the limitations of instrumental and utilitarian thinking, but explain how perception visualises and interacts with the world. From their viewpoint, the intrinsic and subliminal dimensions of the natural environment are revealed using our powers of perception and then expressed through practical reason. Interpretivists begin their exegesis by explaining how our current practices have ignored how we are related to and shaped by the world we live in. They emphasise feeling and perception as having a valid role in the experience of embodied agents making their way in the world. Taylor continues that when environmentalism and politics rely excessively on adapting the model of the natural sciences to environmental and moral theory they tend to narrow our appreciation of sources of the self. That is, environmental and moral theory remains constrained by the dichotomy that separates mind from the natural environment. Taylor states the issue in the following passage:

> There is a big mistake operating in our culture, a (mis)understanding of what it is to know, which has had dire effects on both theory and practice in a host of domains. To sum it up in a pithy formula, we might say that we (mis)understand knowledge as 'mediational'. In its original form, this emerged in the idea that we grasp external reality through internal representations. Descartes in one of his letters, declared himself 'assure, que je ne puis avoir aucune connaissance de ce qui est hors de moi, que par l'entremise des idees que j'ai eu en moi'.[15] When states of minds correctly and reliably represent what is out there, there is knowledge.[16]

In criticising Descartes' disengaged and mediational picture of the world, Taylor argues that the mind is in the world. We are engaged agents in the world, we are not detached and separate from it. He argues that Descartes' picture was part of the scientific Enlightenment that believed it is possible for humanity to measure and quantify the natural environment. This gave rise to a world view where measurement, procedure and technique capture the essence of the natural or real world. This framework has been applied to the humanities without fully addressing its applicability; Taylor's concern is that scientific approaches to governance might not be the best means to order human differences and affairs. It is for these reasons that his work on the public sphere can be extended to examine how practical reason and intrinsic value are related. The issue is to consider whether there are connections between our moral reasons and the intrinsic value in the natural environment, which is discussed in Chapter 4.

I argue that one of the aims of Taylor's interpretivism is to connect our moral reasoning with nature's intrinsic value. The investigation is relevant to how to manage the natural environment. Is it a simple commodity to be managed and used? Here Taylor argues that moral theory involves a fundamental reformulation of humanity's practical reason. Taylor argues that practical reason cannot be reduced to instrumental or procedural forms of reason because values of significance cannot be managed by such criteria. This is because in separating moral and natural properties we become detached from our full resources which include language, perception and spontaneity. In this way, our moral principles are about our place in the natural environment.[17] We have trouble seeing the connection between our moral thinking and instrumental reasoning. It becomes so self-evident that practical reason is simply instrumental in its construction, but our societies fail to see that it is one possible conception among others. 'The moral order in Western modernity becomes associated with the market economy, the public sphere, the self-governing people, among others.'[18] The relationships between environmental politics and social imaginaries are the focus of Chapter 4.

It is important to remember that the distinction between the forms of explanation appropriate to the natural and the social sciences involves our moral reasoning. As rightly noted by Ruth Abbey and Nicholas Smith,[19] the true source of Taylor's anti-naturalism is an ontological claim about human agency as uniquely self-interpreting which challenges the hegemony of the naturalist, positivist and scientific approach to factors of significance for people. Taylor states:

The confidence that underlies this whole operation is that certainty is something we can generate for ourselves, by ordering our thoughts correctly – according to clear and distinct connections. This confidence is in a sense independent of the positive outcome of Descartes's argument to the existence of a veracious God, the guarantor of our science. The very fact of reflexive clarity is bound to improve our epistemic position, as long as knowledge is understood representationally. Even if we couldn't prove that *the malingénie* doesn't exist, Descartes would still be in a better position than the rest of us unreflecting minds, because he would have measured the full degree of uncertainty that hangs over all our beliefs about the world, and clearly separated out our undeniable belief in ourselves.[20]

Taylor's challenge to positivist and procedural social science research, therefore, involves coming to grips with the stronger-order goods that frame and impact on our identity. These are the goods of significance which reflect culture, environment and family relationships; more particularly, these values cannot be reduced to economic calculations and reflect his call to understand the human condition. In this regard, his interpretivism illustrates how positivism and proceduralism can easily collapse into a form of instrumental reasoning, and ends up supporting the accumulation of capital while submerging other values. That is, capitalism and free-market approaches 'just degenerate into organised egoism, a capitulation before the demands of our lower nature'.[21]

Taylor's critics claim that his view of utilitarianism is over-simplified and out-of-date. His answer is to recognise that 'except for extreme "hard line" utilitarians', it is correct to note that most of the proceduralists under analysis do have a place for notions of the good life, which is discussed in more detail in Chapter 5.[22] Indeed, liberals, positivists and utilitarians can be moral realists but they foreshorten the scope of moral philosophy and pay little attention to the good because they focus on the principles and procedures. Taylor's concern is that procedural approaches end up offering bureaucratic and instrumental principles to order societal conduct which submerges awareness of the natural environment. Following Taylor, without an adequate critique and definition of how positivism gives rise to a foundational approach we are unlikely to create successful reform strategies. While foundationalism claims to provide a rigorous and scientific discipline that checks the credentials of all truth claims according to a scientific testing procedure it does not explain how these ideas came to prominence.

3.3 Interpretivism, perception and engaged agency

For present purposes, Taylor's interpretivism challenges the procedural trends outlined above, noting that of all our social imaginaries the most suspect is that which considers authenticity to be the maximisation of consumption and pleasure. Thus, the great strength of Taylor's argument is also that it (re)emphasises the richer moral background within social imaginaries which have been submerged in the rise of individualism and instrumentalism. Taylor notes:

> When we turn to the classic critiques of epistemology, we find that they have, in fact, mostly been attuned to this interpenetration of the scientific and the moral. Hegel, in his celebrated attack on this tradition in the introduction to the *Phenomenology of Spirit*, speaks of a 'fear of error' that 'reveals itself rather as fear of the truth. 'He goes on to show how this stance is bound up with a certain aspiration to individuality and separatedness, refusing what he sees as the "truth" of subject-object identity. Heidegger notoriously treats the rise of the modern epistemological standpoint as a stage in the development of a stance of domination to the world, which culminates in contemporary technological society. Merleau-Ponty draws more explicitly political connections and clarifies the alternative notion of freedom that arises from the critique of empiricism and intellectualism.'[23]

The purpose of Taylor's analysis is not merely to underline the symptoms of a community threatened by individualistic and instrumental research. The great strength of Taylor's argument is also that it (re)emphasises the richer moral background from which the modern stress on individualistic values and instrumental reason took its rise. But before exploring Taylor's relevance to this question it is necessary and important to set the scene by defining Taylor's notion of engaged agency referred to earlier. This involves his metaphysical account of how human experience is influenced by values that are external to us. In his work *The Validity of Transcendental Arguments* he challenges modern Cartesian and Humean accounts of the world that emphasise one side of our experience, that of human subjectivity. He explains:

> This is a conception of the subject as essentially an embodied agent, engaged with the world. In saying that the subject is essentially embodied, we are not just saying that his being a subject is causally dependent on certain bodily features: for instance, that one couldn't see if the eyes were covered, or think if one were under severe bodily stress, or be conscious at all if the brain were damaged. The thesis is not concerned with such empirically obvious truisms.[24]

Here Taylor is following Merleau-Ponty in explaining how our perceptual taking in of the world involves our total engagement with the external world. This is an indirect method of argumentation to overcome modern philosophical dualisms that have given rise to the division between our minds and the world. Moreover, this thinking about the embodied agent involves building on those frameworks which assume clear distinctions between the human and natural environments. In his work on Kant, Taylor emphasises not only the embodied agency but also how language shapes our understanding of human agency. This is an important element in Taylor's debates with the communicative thinker, Jürgen Habermas. The latter theorist moves from Kant to Hegel and back again. In doing so, Habermas emphasises the procedural role of language and creates a perspective that contrasts with Taylor's interpretivism. This debate is highly relevant to environmental politics in the search to relate humanity with the intrinsic value in the natural environment, which is a focus of Chapter 7.

However, for Taylor interpretivism is about recognising the plurality of values that exist in modern communities and which cannot be reduced to utilitarian calculi. Here the work of Jonathan Blakely provides a fascinating summary of Taylor's interpretivism while also using the works of Abbey and Smith. All three authors share the view that Taylor follows Heidegger in 'ontologising' hermeneutics into the very structure of human action. It is well-known that prior to Heidegger, hermeneutics was concerned primarily with texts. But Heidegger advanced the view in *Being and Time* that humans are embodiments of self-interpreted meaning. Blakely writes:

> Like an octopus's ability to change colors or the phases of a butterfly's life, self-understandings change and restructure the very way humans exist in the world. So Heidegger wrote that in the case of human beings 'understanding constitutes this being.[25]

Blakely argues that Taylor's interpretivism involves processes of self-understanding where self-interpretation itself reflects our humanity. For Blakely, it is human moods which are of critical importance in orienting us in the world. The self is not some automaton devoid of colour and passion as epitomised by some economic and procedural approaches to the question what is human? Disclosure and moods are a central feature of interpretivism and are revealed through our self-understanding and interpretation of our humanity.

For the interpretivist it is important to recognise the limitations of foundational, positivist and procedural approaches, where self-under-standing involves 'attunement' or 'moods' as part of the process of discovery. Blakely argues that Taylor's interpretivism emanates from Heidegger's view that human beings are 'always already in a mood' such that they 'never master a mood by being free of a mood, but always through a counter mood' be it 'elevated' or 'a bad mood'.[26] These various states of our humanity are encapsulated in Taylor's interpretivism, which emphasises the notion of a person as a self-interpreting animal capable of making strong evaluations. It is argued that modern political theory has lost sight of the values that make the world, values we ignore when we rely on neutral conceptual and scientific frameworks. For Taylor, political discussions concerning nature's voice and its intrinsic value have become submerged by procedure and technique in the construction of the public spheres of modern liberal democracies. Taylor explains:

> Whereas for the simple weigher what is at stake is the desirability of different consummations, those defined by his *de facto* desires, for the strong evaluator reflection also examines the different possible modes of being of the agent ... Whereas a reflection about what we feel like more, which is all the simple weigher can do in assessing motivations, keeps us as it were at the periphery; a reflection on the kind of beings we are takes us to the centre of our existence as agents ... It is in this sense deeper.[27]

The notion of strong evaluations provides an important link for environmental political research as it suggests that some goods define and frame other goods. The obvious link is that the natural environment frames the freedoms we enjoy. From this perspective, a problem with naturalism and positivism is that they overplay their hand in relying on the procedures of scientific testing to control and model the natural environment. Here the relevance of Taylor's interpretivism involves recognising the connections between the environmental and moral dilemmas that confront us. Clearly, this awareness means that these dilemmas cannot necessarily be reconciled by economic, procedural or utilitarian approaches. When we engage in interpretation we recognise the fragility of goodness and the environmental and moral dilemmas that exist between us.

In this regard, interpretivism is used to explore and reveal the various social imaginaries that gain sway in modern communities. For example,

Taylor notes that interpretivism may be used to explore the rise of positivism and the force that this ideology commands in academic research and in turn environmentalism and public policy. Taylor's work on interpretivism is part of his evaluative framework which searches for meaning and intrinsic value in the natural environment. But before exploring Taylor's relevance to this question it is necessary and important to set the scene by defining Taylor's notion of engaged agency referred to earlier. This involves his metaphysical account of how human experience is influenced by external values in the world which is the focus of the chapter that follows.

3.4 Conclusion

Taylor's evaluative and interpretivist framework offers a different way to think about what is significant in human communities; it enables us to move away from the tendency of positivist research to be beholden to free markets. He proposes an evaluative and interpretivist framework based on the argument that we can never fully predict human behaviour as modern social science assumes. Therefore, he proposes an account of the person or self that he describes as 'expressivist' or 'self-interpreting'. The self is constituted, at least in part, through its own self-interpretations. These self-interpretations are not something that can be understood with certainty and thus require sensitive analysis which requires an ongoing conversation which continually questions 'our subsequent reinterpretations'.[28]

Interpretation, interpretivism, sensitivity and judgment are integral to understanding our human communities and are not something that can be modelled with abstract neatness. Taylor's evaluative framework can be used to foster an environmental research agenda that does not only focus on moral values, as important as they are, but also provides an interpretivist analysis of meaning. The evaluative framework must be developed to supplement the mainstream's reliance on an instrumental social imaginary which emphasises procedure at the expense of interpretation. Mainstream environmental and social research rely on a dominant framework which serves the needs of decision-makers without fully recognising differences between 'individuals' or 'selves'. Taylor emphasises the notion of the self as capable of individual reflection and re-evaluation which is a way of thinking foreign to the modern social science preference for instrumental reason and procedure.

4
Taylor's Interpretivism, Social Imaginaries and the Natural Environment

This chapter examines Taylor's social imaginaries and his work on the natural environment. I extend this approach to examine Taylor's engaged agency thereby putting us in touch with the natural environment.

4.1 Introduction

Over the past 20 years Taylor has developed the notion of the social imaginary to illustrate how social structures impact communities and the natural environment. This chapter explores Taylor's recent work on the social imaginary using his metaphysical analysis to deepen our interpretation of Western modernity and how it defines what the natural environment is. To achieve this understanding this chapter explores both the original and contemporary issues concerning the dilemmas of unlimited economic growth and the preservation of the natural environment. These dilemmas can be traced back to debates concerning the intrinsic value of the natural environment which can also be found in Taylor's work on the role and validity of transcendental arguments.

Taylor's interpretivist framework is developed in this chapter to explore the metaphysical viewpoint that the natural environment contains intrinsic value and meaning. The interpretivist framework is implicit in Taylor's *Modern Social Imaginaries*[1] where he aims to uncover the direction and patterns of meaning in the natural environment and in contemporary social structures.[2] His work on social imaginaries is relevant to ecological politics, as he rethinks our relationships with the natural environment. Since the publication of *The Brundtland Report* in 1984 the question of our relationship with the natural environment has become part of mainstream economic and political enquiry. Yet Taylor was already discussing these issues in his early communitarian and interpretivist work when he explored the notion of steady-state social structures. In more recent work his interpretivist framework examines the major political impediment to the consideration of steady-state social arrangements, namely 'intolerable inequality that is made "provisionally tolerable" only by rapid economic growth'.[3]

The works of Dreyfus, Gadamer,[4] Heidegger,[5] Nussbaum[6] and Taylor[7] are used as representative of interpretation. In the four sections following this introduction this chapter examines the concept of the social imaginary as well as the buffered self in environmental contexts. Section 4.2 defines Taylor's work on modern social imaginaries to offer environmental direction. Section 4.3 examines how our social imaginaries influence us as buffered selves in the natural environment. Section 4.4 develops interpretivism to explore nature's meaning and intrinsic value. It uses recent work on secularisation with environmental politics in mind. The chapter concludes in Section 4.5 with some comments about Taylor's relevance to environmental approaches that focus on intrinsic value.

4.2   The social imaginary, nature and the Enlightenment

Having sketched some of the key features of Taylor's interpretivism and metaphysics it is necessary to relate it to his work on social imaginaries, which explains how the social system holds these values together. He explains the social imaginary in the following way:

> By social imaginary, I mean something much broader and deeper than the intellectual schemes people may entertain when they think about social reality in a disengaged mode. I am thinking, rather, of the ways people imagine their social existence, how they fit together with others, how things go on between them and their fellows, the expectations that are normally met, and the deeper normative notions and images that underlie their expectations.[8]

Taylor's recent work, *Modern Social Imaginaries*, is about engaging the will of the people through collective social imaginings. The work on imaginaries challenges the ideas and theories that have been used to consider our place in the world. He argues that our ways to think about the natural environment have been under-theorised since Hume's *Enquiry* and Kant's *Critique of Pure Reason*.

Taylor's interpretivist analysis of the social imaginary is designed to examine the foundational belief that philosophical and political theories can correctly outline our beliefs, connections and relationships with the natural environment. Western modernity is inseparable from a certain kind of social imaginary that involves the contingencies and significantly defined forces that influence our attitude towards the natural environment. This supposition involves combining the multiple modernities to incorporate divergent social imaginaries which shape modern social structures. The idea of the social imaginary, therefore, can be used to explain how environmental politics is involved in often divergent and complex interpretations to make sense of the natural environment. As Taylor has observed:

> Western modernity on this view is inseparable from a certain kind of social imaginary, and the differences among today's multiple modernities are understood in terms of the divergent social imaginaries involved. This approach is not the same as one that might focus on the *ideas* as against the *institutions* of modernity. The social imaginary is not a set of ideas; rather it is what enables, through making sense of, the practices of a society.[9]

Taylor uses the ideal of the social imaginary to make sense of the idea that the natural environment contains meaning and value. He does this by explaining how the modern scientific view has dominated our conception of the natural environment under the rubric of natural law. Yet, his critics claim that his enchanted view of the natural environment makes our relationships with the natural environment seem mysterious. However, the implication from the enchanted view is that the natural environment contains conceptualised value that people are then able to articulate and express. For reasons such as these he uses the idea of a social imaginary to explore how we make sense of the natural environment and, obviously, to explain the patterns of interaction within cultures and modern communities.

From this perspective, a key aim of Taylor's work is to explain how to understand and make sense of the values that we perceive in the natural environment. Of course, Taylor is not attacking scientific method here because science has been successful in explaining the laws of nature and its physical processes. These are major achievements of the modern era. Nevertheless, with the rise of the scientific method the disenchanted viewpoint seems so obvious when we think of intrinsic value and social values only as emanating from thinking minds. It is through the notion of the social imaginary that we begin to explore the existential and metaphysical dimensions of these relationships. The imaginary challenges modern Humeans and Kantians to reconsider how knowledge of the natural environment is constructed; Taylor reminds us that, in one sense only, Kant is downstream from Hume in suggesting a distinction between analytic and synthetic forms of knowledge.[10] This philosophical separation between mind and world has dominated modern political and philosophical debate since the Enlightenment.

4.3 Social Imaginaries, Buffered Selves and the Natural Environment.

Social imaginaries and interpretation are the key frameworks that are relevant to our search for meaning and intrinsic value in the natural environment. For example, Taylor's *Sources of the Self* reflects a narrative structure that has only recently become explicit in his analysis of the processes of modern secular societies.[11] In his recent work, *A Secular Age*, he explains the narrative distinctions between enchanted and disenchanted visions of the natural environment. For the disenchanted vision of the natural environment, the idea is that moral meanings and intrinsic value are not found in it.

However, by way of contrast, in the enchanted view values can exist independently of people.[12] These values are revealed through language which co-discloses different meanings and values. Taylor's *Modern Social Imaginaries* was written before *A Secular Age* and can be used to understand his master narrative that the natural environment cannot be reduced to disenchanted interpretations. The master narrative guiding his work is designed to offer an account of the factors that have given rise to the Western Enlightenment. As we shall see throughout this chapter, interpretivists such as Taylor develop aspects of Heidegger's and Merleau-Ponty's work on our basic ability to move in the natural environment (world) and move beyond the search for universal rules and procedures. Taylor's interpretivist approach considers the common purposes that are a product of democratic interchange and deliberation in civilising society.

Taylor's developments in these areas of philosophy explore the intrinsic value of the natural environment through the 'space of reasons' connecting humanity with it.[13] This analysis of the processes that have led to our current procedural and secular age offers a series of political steps to rethink environmental values and political structures. One first step involves exploring not only our representations of the world external to us, but also our unreflective capabilities and capacities that allow us to get about in the natural environment. Here Taylor argues that the commonalities with the natural environment can be seen as a feature of the imaginary, revealing the idea that humanity has the means to come in contact with being, language and our relationships with the natural environment. This line of thinking reflects the sceptical dimension in Taylor's critical engagement with the social imaginary – this involves his dissatisfaction with the inwardness associated with the logic of modernity. His aim is to develop the view that the enchanted world was one in which these forces (cosmic, embodied or spiritual) could cross a porous boundary and shape our lives. He argues that these forces influence our psychic and physical well-being. He continues that one of the big differences between us and them is that we live with a much firmer sense of the boundary between self and other:

> We are 'buffered' selves. We have changed. We sometimes find it hard to be frightened the way they were, and indeed, we tend to invoke the uncanny things they feared with a pleasurable frission, sitting through films about witches and sorcerers, they would have found this incomprehensible.[14]

That is, re-enchantment does not undo the disenchantment associated with modernity's social imaginary. Taylor asks us to see the contrast with a modern person who is feeling depressed, associated as it is with melancholy. Taylor explains the contrast in the following way:

> [T]hat for the modern it is 'just your body chemistry, you're hungry, or there is a hormone malfunction, or whatever'. The modern feels relieved and the 'feeling [which] is ipso facto declared not justified'. Accordingly there is no real meaning here, it is just the way things feel and can be explained by a causal action utterly unrelated to the meanings of things.[15]

Clearly Taylor's analysis of the disengaged stance depends on the distinction between mind and world. This is where the physical dimension is relegated to being 'just' a contingent cause of the psychic. Taylor continues that a pre-modern may not be helped 'by learning that his mood comes from black bile, because this doesn't permit a distancing'.[16] Black bile *is* melancholy. Taylor says that the pre-modern just knows that he's in the grips of the real thing.[17] The point of this rather difficult example is to reveal the differences between pre-modern and modern approaches to physical and psychic approaches to value. That is, it is important to remember how meaning can emerge from outside our thinking minds.

4.4 Developing interpretivism, nature's meaning and intrinsic value

Taylor's alternative to closed world scientific thinking involves his tendency not to attribute external value independent of valuing subjects. And so he explores the non-arbitrary and projective relationships that place demands on people. These demands are revealed through our being-in-the-world. He explains:

> So 're-enchantment' in this sense doesn't undo the 'disenchantment' which occurs in the modern period. It re-establishes the non-arbitrary, non-projective character of certain demands on us, which are firmly anchored in our being-in-the-world. These demands do not just emanate from us, but at the same time, they are not inscribed in the universe (or the Universe and God) independently of us. They arise in our world.[18]

Taylor argues that we must move beyond projectivism to understand our more basic relationships with the natural environment. Taylor is clearly challenging the position that reasons are external projections from us. He has stated that he finds these ideas in David Hume's work. Hume's work attributes a disengaged stance to the natural environment. Taylor argues that this process of reasoning sets in motion social trends that lead to an excessive focus on accumulation, appetite and economic growth. The modernist approach resembles Hume's scepticism that certainty emanates from our minds. This leads to the view that environmental and social issues can be solved by our economic and market mechanisms.[19]

Rather, these environmental relationships have to be seen as a fact of *experience*, not as a matter of theory or belief. They are revealed in our direct engagements with the world thereby providing a base to develop environmental awareness. The challenge for environmental politics, therefore, is to directly engage the will of the people rather than relying on market mechanisms to create improved outcomes. His first response to the environmental and social problems of modernity therefore involves the argument that these dominant frameworks assume moral judgments that are universal and incontestable. A standard criticism that is advanced in his work is that universal solutions have the propensity to ignore local values in the search for what is universal and incontestable.

Here, Taylor uses ideas from Hegel and Heidegger to nurture a synthesis of humanity and nature; namely, a synthesis that includes the natural environment as a background value requiring political deliberation in dealing with our environmental dilemmas.[20] In this way we begin to understand our existence but also other entities in the natural environment. We begin to develop capacities, capabilities and direct engagements which open us to the mysteries of the natural environment. This interpretivist way of thinking is discussed more fully in Chapter 5.[21] Furthermore, the pervasive influence of Hume and Kant has changed Western thinking, but there has been limited examination of our more basic coping and perception skills as they extend to the natural environment. Our coping abilities are part of our unreflective actions which are important when we consider how we might create better ways to interact with the natural environment, which is discussed in Chapter 5.

From Kant's critique of Hume, it is well-known that analytic knowledge is derived from internal thinking and creates a clear distinction from the world, while synthetic knowledge is taken from sense experience as people interact with the world. Often synthetic knowledge is associated with the laws of nature as studied by the natural sciences. Taylor examines these Kantian distinctions to discuss how they shaped philosophy's understanding of modernity together with their influence on how to view the natural environment. Obviously, analytic propositions are true by virtue of their meaning, while synthetic propositions are true by how their meaning relates to the world. But as we have seen Taylor goes further to examine our more basic interactions in the natural environment and as I have argued takes philosophy in a way that directly engages with the natural environment.

Taylor developed the idea of social imaginary together with his metaphysics to explore the distinction between synthetic and analytic approaches to knowledge. His aim is to interrogate the distinction as it bears on people's self-interpretations and understandings of the natural environment. In utilising this framework, I investigate the hypothesis that it is possible to understand our relationships with the natural environment when we consider the dominant logic of modernity and achieve clarity concerning its processes. These processes, shaped as they are by modern epistemology, explain how our interactions with the natural environment reflect the previously mentioned distinctions between what contingently changes in the world and the existing facts that shape our being-in-the-world.

Here philosopher Iain Thompson has pointed out that Taylor's analysis focuses on closed world structures but tends to overlook the sources within our own cultures. The implication of Thompson's argument is that Taylor's interpretivism and metaphysics cannot prove external and intrinsic value. Thompson argues:

> What *A Secular Age* most forcefully objects to about the post-Nietzschean proponents of wholesale secularization is that their philosophical worldviews take the shape of what Taylor calls a 'closed immanent frame' (p. 548) or 'closed world structure' (p. 589), where (in both cases) 'closed' means 'closed to transcendence' (p. 557). Of course, what exactly 'transcendence' means is not obvious, and Taylor himself exploits this fact in his efforts to challenge both the cogency and the desirability of the post-Nietzschean quest for a secularization without remainder.[22]

Thompson maintains that the aim of *A Secular Age* is to explain *secularisation*, that is, 'the rise of unbelief'. He argues that *A Secular Age* is dedicated not just to explaining but also to *resisting* 'the coming of a secular age'.[23] Thompson is implicitly questioning what Taylor means by transcendence; but, Taylor responds that this is really an indirect question to exclusive humanists and closed world thinkers. The question that animates Taylor's whole body of work is this question: 'is this all that there is?'

Yet critics such as Thompson do not fully explore Taylor's transcendental work on our direct coping which opens us to the possibility that there is meaning and intrinsic value in the natural environment. Moreover, in Taylor's work on how language expresses these values he points out that the meaning and value exists in nature influencing and shaping our lexicon, which is discussed more fully in Chapter 7. But for present purposes, our language is influenced by nature's external and intrinsic value. For example, language and perception have a two-stage approach which puts us in the natural environment. The first stage of Taylor's strategy is to break down the dualism between the mind and the natural environment so that the environmental values in our world also influence and are influenced by language. The Taylorian approach involves the possibility that there is meaning and intrinsic value in the natural environment. His second stage is to explain that external values and intrinsic value which shape our language. This ability through language provides humanity with the means to come in contact with external and intrinsic value that are external to the thinking mind.

I have argued that Taylor uses these arguments to challenge the dominant representational view of human behaviour which assumes the priority of our inner thoughts and ends up devaluing discussions concerning intrinsic and theistic values. The relationship between our thoughts and their relationship with the world have also been examined by important theorists such as Stephen K. White 'who draws a distinction between fullness as conceived by Taylor and the search for dearth experiences'.[24]

However, for White 'dearth' experiences involve our search for fulfilment in our lives. For White, dearth experiences overlap with Taylor's transcendental analysis, but it is not totally subsumable within it, particularly in the modern world.

Here White's work on the phenomenon of dearth draws attention to those experiences which are not dependent on a transcendent source that defines one's highest good. White argues:

> Although being 'born again' is probably the most frequently identified depth experience in America today, it is crucial not to let that one type be mistaken for the entire category of experience in which I am interested. I take being 'born again' to be an instance of what Charles Taylor (2007, 5ff) calls an experience of 'fullness'. I would differentiate that general mode from another I will call 'dearth'. These two modes comprise the category of depth experience. My underlying aim is to consider the implications such experience can, and ought to, have in the life of a late-modern democratic society. Before turning to that task, it is necessary to further elucidate 'fullness' and 'dearth'.[25]

For White it is especially important to see that neither fullness nor dearth needs to be restricted to its most extreme manifestations. A dearth experience is about exploring what is missing in our lives and for White might be found in our existing social space. Thus, reconceived by White, a dearth (depth) experience is a phenomenon that political reflection should engage rather than try to avoid entirely. Toward that end, White speculates about how moderately:

> [W]ell-off citizens of Western democracies might cultivate such experience in ways that honour admirable values in our tradition of democratic thought, while using the extraordinary affective force of fullness and dearth to dislocate us somewhat when we become complacent about those values.[26]

For White it is important to examine not only those sources that empower meaning but also how dearth reflects that loss of meaning whether in the form of death or suffering. For the purposes of my argument, however, a search for dearth experiences does not fully engage with the sense of intrinsic value and mystery that involves human fullness. Fullness, for Taylor 'is a condition and an outcome of the recognition of, and gesture towards, transcendence'.[27]

For the purposes of my argument, Taylor offers an indirect transcendental argument that goes beyond the buffered and inward self. Life-changing fullness is not only beyond ordinariness, but also beyond the self, and embraces sacredness [transcendence]. Perhaps the clearest analysis of Taylor's work on metaphysics and transcendence has been provided by the philosopher, John Rundell. He argues sacredness and transcendence is opening oneself to mystery, depth and verticality. This metaphysical strategy was outlined in previous chapters of this current book to explore the intrinsic value of the natural environment. Rundell's analysis develops the quest to achieve 'a break-out from the immanent frame' which amounts to a paradigm shift in our collective understanding of modernity.[28] Following Rundell it is important to remember that:

Taylor argues that this has occurred from two directions which has altered and 'transformed our understanding beyond the usual scope of the ordinary, either within or outside its embeddedness in liberal civilisation...Taylor draws on the work of the French poet and worker's activist of the early twentieth century, Charles Péguy'.[29]

From these poetic sources Taylor's hermeneutics of creative renewal or conversion involves the following four key features as developed by Rundell. First, 'authentic action links ordinary, everyday life, present and past together aligning them with transcendent or cosmological time'.[30] Second, a 'plurality of mystical experiences in which all of Judaism and Christianity contribute their own particular versions of mystery, and their access to it'.[31] Third, there is an emphasis on the image of harmonious cohesion and integration along the lines put forward by Péguy, Durkheim and Mauss in their defence of modern corporatism'.[32] And fourthly, 'the polytheism of sacred practices involves a universalistic attitude towards salvation which is available to everyone: there is no space of Hell, no space of banishing the negative to the outside'.[33] Rundell opines that we are outside the paradigm of the self-defining subject and have come to a dimension almost entirely beyond ourselves.

Therefore it is a categorical error to focus on religious subjectivism because Taylor's hermeneutics combines the poetics of Romanticism and religious experiences to create awareness of cultural, environmental and significant values. Taylor's developments in hermeneutics and interpretivism therefore 'involve the counter-paradigm or developing a counter-imaginary to modernity'.[34] I have used interpretivism and the counter-imaginary to put people in touch with the external and natural environment. Rundell's analysis goes further to illustrate how transcendence itself locates a space outside of the self-defining subject. He locates fullness and human flourishing in a dimension that is not fully addressed by depth and dearth theorists such as Thomson and White. This transcendental dimension, I contend, offers means to move beyond closed world thinking and explain that within modernity, commonality, metaphysics and transcendence seem only illegitimately possible in our instrumental, plural and secular world.

It is therefore important to bear in mind that the idea of a cosmological and social imaginary does not provide the same explanations that follow from procedural and reductionist accounts of modernity. The procedural approach to politics, moreover, focuses on the ideas and theories that have shaped current understanding and vision of the natural environment. In these additional ways, the idea of cosmological and social imaginaries offer a contrast to those procedural approaches that interpret the practices internal to a society, which have led to particular instrumental way to visualise the natural environment. It is for these reasons that Taylor argues that such values within the cosmological imaginary must be explored in the public sphere, which is the focus of Chapter 6.[35]

4.5   Conclusion

In conclusion this chapter has argued that it is important to consider Taylor's work on the cosmological and social imaginary if we are to appreciate the intrinsic value of nature. I argued that our place in the natural environment involved hermeneutic and metaphysical concepts implicit in Taylor's evaluative approach. These strands of thought combine with his work on engaged agency, language and perception to put us in touch with external and intrinsic values.

Since the Western Enlightenment, humanity has acquired the technological means to transform the natural environment. However, a full philosophical and political analysis of the destructive tendencies within modernity has not been undertaken using Taylor's work. While contemporary environmental politics has begun to integrate the natural environment into its social imaginary, the economic features of the imaginary have been given prominence. Taylor's work on the social imaginary identifies problems with neo-classical economic approaches and modern approaches that assume humanity can control and transform the natural environment. Taylor's solution is to explore the social imaginary to reveal the logic in modern meta-narratives and repair damaged environmental and social relationships. This involves understanding the connections between ecosystems, and economic and social development using our cosmological and social imaginaries. These issues are developed in the next three chapters.

5
Taylor's Metaphysics, Merleau-Ponty and the Natural Environment

Taylor has stated that he concurs with aspects of Deep Ecology through his evaluative and interpretivist framework. I use his work on Maurice Merleau-Ponty's direct coping and perception to explain this affinity with Deep Ecology. This involves exploring the meaning and value provided by nature which involves how we are open to a world which can be explored, learnt and then theorised about.

5.1 Introduction

For many years, environmental ethics has been concerned with exploring the intrinsic value and meaning in the natural environment. Taylor has stated that he concurs with aspects of Deep Ecology through his evaluative and interpretivist analysis of our place in the world. For the purposes of environmental awareness, one of his principal aims is to explain how nature's meaning and value influences our being-in-the-world. He undertakes this task by deconstructing the Enlightenment's emphasis on instrumental and technical approaches to explaining the meaning and things in the natural world.[1]

Taylor introduces us to an environmental conversation that challenges the dominance associated with a type of exclusive humanism which is associated with closed world thinking – closed world thinking is connected with the methods of the natural sciences. Taylor challenges the application of this naturalistic approach to human and environmental issues and argues that it leads to an arid and technical vision of nature. Moreover, the dominant presuppositions of the natural sciences assume that it is possible to control and model the processes of nature. Taylor challenges these presuppositions by revealing how instrumental approaches deny the belief that nature contains meaning that can be disclosed. It is therefore the task of interpretivism to provide arguments and conversations that disclose these meanings and values.

Through the auspices of Taylor's analysis, my examination adds to the work of environmental theorists such as J. Baird Callicott,[2] Aldo Leopold[3] and Holmes Rolston III.[4] The addition to their environmental work involves asking how our basic engaged coping opens us to the possibility that an enchanted world exists for interlocutors receptive to the suggestion that meaning exists in the world.[5] I argue that a re-enchanted vision of nature emerges through this interpretivist framework which begins by questioning how we actually experience things in the world such as nature. In Taylor's view, our awareness and interaction in the natural environment is better understood through an approach which acknowledges our existence against a background of significant features associated with particular environmental and social contexts.

The relevance of Taylor to environmental ethics involves his work on how we actually interact and interpret the world. Briefly, his first step is to examine whether we visualise the natural world simply through the internal representations we create in our mind. The next step is to explore how we are directly engaged with the world which leads to the notion that we are not simply beings with a consciousness but that we are subjects who are situated in the world. These steps challenge the modern epistemological view that we interact with the world only through the representations our minds create. These interactions and relationships must be understood in very different terms to those developed in the natural sciences which involve our primordial engagement with the world. That is, our situation in the world involves grasping the primordiality of object engagement which itself leads us to reject the view that we can model the objects and things in the world with certainty.

I undertake this task in the following four sections of this chapter by showing that Taylor's interpretivism is linked to his metaphysics and builds on insights from Merleau-Ponty's work on engaged agency. Section 5.2 introduces Taylor's work on Merleau-Ponty's direct engaged approach to put us directly in touch with the world. Section 5.3 develops connections between engaged agency, self-understanding and nature's intrinsic value. Section 5.4 uses Merleau-Ponty's framework to explore how Taylor develops our primordial coping abilities and language. Section 5.5 concludes the chapter by exploring some environmental connections among perception, language and strong evaluations.

5.2   Taylor's work on Merleau-Ponty, metaphysics and nature

This section outlines how Taylor follows Merleau-Ponty in explaining that the most basic form of our existence appears as a body-subject. Of course, Taylor is not asserting that nature contains meaning and intrinsic value independent of any valuing subject, but he is offering interpretivist thinking to reveal the limitations of our instrumental and scientific presuppositions.[6] This interpretivist approach therefore offers an indirect argumentative strategy to come in contact with how we experience these meanings and values. To do this Taylor draws on Merleau-Ponty's analysis of our engaged agency in the world, as well as on ideas from Herder, Humboldt and Heidegger. On the view developed in this chapter, we are not simply consciousness in an objective body but we are actually a different kind of being.

In Taylor's classic work 'Overcoming Epistemology' he continues to challenge modern accounts of our place in the world which emphasise one side of our experience, that of our human subjectivity.[7] Here it must be stressed that Merleau-Ponty explains the role our perceptions play in absorbing and reflecting the meaning and values in the world. Perception involves our total engagement with the external and natural world. I argue that this Taylorian strategy is based on the fact that we must begin to distinguish how we experience things in the world from merely instrumental and subjective approaches. For otherwise we would have experience which is not experienced as being of anything; it would be an experience without an object, and this we can see to be an impossibility.[8] For Taylor, naturalistic and subjective contributions to interpreting meaning and things in the world leave us reliant on economic, instrumental and technical assumptions. In reality we are beings who exist in natural and social environments which leads to the view that we are not simply consciousness in an objectified body.

Taylor emphasises Merleau-Ponty's notion of direct engagement with the world that we experience as body-subjects. Taylor's metaphysics involves 'a living attempt to deploy an argument of transcendental type, and hence as a continuation, at least in one sense, of the enterprise Kant started'.[9] In undertaking this task Taylor is implicitly challenging the dominant view in the social sciences that we gain access to the world only through our consciousness. Here Taylor traces this line of thinking back to the work of Kant[10] which held that all knowledge involves the co-operation of the mind.[11] This mind dependent approach to our place in the world reflects the view that there can be anything like experience or awareness of the world in the first place. Here Taylor offers a further challenge to those Kantian views which focus on experience being processed solely through thinking minds. Kant's view offers a representational view of the world. Taylor challenges the representational view by revealing how it is dependent on the primordiality of object engagement which was discussed earlier in this chapter. Our primordial grasping of objects and things in the world dissolves temptations to support the tenets of anti-realism which suggests that there is no reality or value in the world.

As stated earlier, Taylor is essentially challenging Kant's view that only through our thinking minds can we have experiences. This is associated with a representational view of the world. Accordingly, when we assume that we experience the world only through our concepts and categories we close ourselves to the otherness of nature. We lose sight of the potential meaning and value that is contained in nature. Furthermore, Taylor points out that Merleau-Ponty gives emphasis to our senses and perceptions which are integral means through which we directly engage with the natural environment. Taylor argues that Merleau-Ponty's approach gives rise to a direct access approach which is 'the foundation of all other ways of having a world'. That is, our basic abilities in perception differentiate us from other animals. He emphasises that our perception occurs through our total comportment in the fields we operate in. He explains that our 'perception of the world is essentially that of an embodied agent' – Taylor emphasises the term 'essentially' to express how perception depends causally on certain states of our bodies: he points out that this thinking is not the same as the assumption that one could not see if my eyes were not in good condition. He continues:

> The claim is rather that our perception as an experience is such that it could only be that of an embodied agent engaged with the world. Let's consider. Our perceptual field has an orientational structure, a foreground and a background, an up and down. And it must have; that is, it can't lose this structure without ceasing to be a perceptual field in the full sense, i.e., our opening onto a world. In those rare moments where we lose orientation, we don't know where we are; and we don't know where or what things are either; we lose the thread of the world, and our perceptual field is no longer our access to the world, but rather the confused debris into which our normal grasp on things crumbles.[12]

For Taylor, the key to engaged coping is that our consciousness is parasitic on perception such that how we encounter things in the world actually gives rise to our awareness of nature's environmental meaning and intrinsic value. This approach offers a means to explore nature's meaning and value leading to the view that the external world is not simply empty and devoid of intrinsic value. Taylor is challenging the naturalist view of the world that offers a disengaged view of the self. This view is based on the claim that the human self can be defined using the methods of naturalism. Furthermore the tenets of naturalism dominate the social sciences. But the analysis of our engaged agency involves exploring tendencies and trajectories which are implicit within various political approaches.

It will recalled from Section 5.2 that Taylor's analysis begins with Kant speaking simply in discussing experience before moving to Merleau-Ponty's concern to get beyond subjectivistic formulations which focus on thinking minds.[13] Taylor deconstructs this subjectivism which he believes leads to a disenchanted vision of the natural environment. He challenges this focus by considering how we actually acquire knowledge of the natural environment.[14]

The relevant key points involve:

'(i)   The argument for direct engagement with the natural environment begins by critiquing approaches such as those of Kant to whom our a priori sensibilities shape our subjective stance to the world. Taylor challenges Kant's claim that we interact with the world only through the representations we hold in our mind.

(ii)   Taylor's interpretivism emphasizes our direct engagement in the world which leads him to question the primacy of individualism within modern epistemology. Individualism ignores our social embodiment in the practices of our being-in-the-world.

(iii)   Dissolving the assumptions of individualism the focus turns to our being-in-the-world which allows us to refer to key focal activities which Heidegger termed pragmata (earth, sky, mortal and the divine), which are the factors of significance that we encounter directly and cannot be explained in neutral and scientific terms.

(iv)   These elements combine in such a way that we acquire knowledge and then give status to things or objects of significance.'[15]

These four elements of Taylor's interpretivism give rise to the different levels of awareness that are used to explain our place in the natural environment. It must be stressed that Taylor's interpretivist strategy challenges the dominant approaches which focus solely on the mind and our inner representations. Taylor explains the key implications associated with our agency and its connections to the natural environment. He continues:

> We only have knowledge as agents coping with a world, which it makes no sense to doubt since we are dealing with it. There is no priority of the neutral grasp of things over their value. There is no priority over the individual's sense of self over the society; our most primordial identity is as a new player being inducted into an old game. Even if we don't add the fourth stage, and consider something like the divine as part of the inescapable context of human action, the whole sense that it comes as a remote and most fragile inference or addition in a long chain is totally undercut by this overturning of epistemology.[16]

The key point that emerges from this passage involves a way of thinking that leaves space open for external values so that we minimise the potential loss of meaning when we operate in closed world systems. This is Taylor's indirect philosophical approach to realism which I use to deepen the earlier arguments in Chapter 2 in which Taylor's political work suggests the possibility that nature contains meaning and value. He deepens these arguments by using Merleau-Ponty's argument that we are natural beings who are permeated with rationality.[17]

Taylor explains his basic argument for direct engaged coping as involving our grasping of things as primordially one of bodily engagement with them. Accordingly, we begin to understand that 'we are in contact with the reality which surrounds us *at a deeper level than any description or significance – attribution –* we might make of this reality, and that this dissolves the temptations to anti-realism'.[18] This deeper reality allows us to realise that there is more meaning in the world than the one we attribute to it – this discovery does not deny the important insights of science but actually provides a partial reconciliation between science and direct coping. That is, science may help us understand the boundaries and limits on our ability to cope in the world such that we are more responsive to what the universe and nature are.

5.3 Searching for nature's value, engaged agency and nature's value

However it must be stressed that our direct engagement with the natural environment has important ethical implications for how we deal with it. These implications have been explored by environmentalists such as J. Baird Callicott who explain thus:

> That intentional consciousness must be embodied follows from the most basic reflection on perception and experience–which is rooted in sensory organs. No body, no organs of sense. No organs of sense, no intentional consciousness. No intentional consciousness, no experience for phenomenological exploration. Maurice Merleau-Ponty writes: 'When I turn towards perception, and pass from direct perception into thinking about that perception, I reenact it, and find at work in my organs of perception a thinking older than myself of which these organs are merely a trace'.[19]

Taylor concurs with Callicott et al. that when we reflect on our basic coping abilities we find that our customs, dispositions and habits also put us in the natural environment. For some people this might reflect their cultural background's horizons of meaning which reflect their place in nature. The point is that for strong evaluators, nature is a good that cannot be approached solely from a naturalistic perspective. Nature is not a zero-sum game.

Furthermore, implicit in the views of both Callicott et al. and Taylor is the supposition that our basic coping abilities engage us with the world. As such, we merge mind and body. My environmental argument is that by using Taylor's interpretivist and holistic approach, we tap into environmental values such as those espoused by Aldo Leopold:

> Like winds and sunsets, wild things were taken for granted until progress began to do away with them. Now we face the question whether a still higher 'standard of living' is worth its cost in things natural, wild and free. For us of the minority, the opportunity to see geese is more important than television.[20]

Leopold expresses and recognises the various meanings that can be found in the natural environment. Following Leopold and Taylor, the awe and beauty of the natural environment have been submerged in the quest to transform nature for human purposes only. These are important insights for environmental ethics and politics because without intentional consciousness there is no experience of the intrinsic beauty and meaning of the natural environment. These environmental observations, therefore, give us access and insight into different features of the natural environment thereby challenging our dominant instrumental and procedural viewpoints. This suggests a new conception of under-standing nature which puts us in touch with it. Clearly challenging our anthropocentric approach the aim is to explore the values in nature. He explains:

> [T]his is, as it were, a condition which arises even in a disenchanted world: we are unprotected; now not from demons and spirits, but from suffering and evil as we sense it in a raging world. There are unguarded moments when we feel the immense weight of suffering, when we are dragged down by it, or pulled down into despair. Being in contact with war, or famine, or massacre, or pestilence, will press this on us. But beyond suffering there is evil; for instance, the infliction of suffering, the cruelty, fanaticism, joy or laughter at the suffering of the victims. And then, what is worse, the sinking into brutality, the insensible brute violence of the criminal. It's almost like a night-mare. One wants to be protected, separated from this. But it can creep under your guard and assail you, even in a disenchanted world.[21]

Here Taylor explores some of the consequences that might follow from modernity's predilection for a disenchanted world. We no longer are able to protect ourselves from modernity and its predilection for more proce-dure, no longer do we have the internal resources to it. Our resources to respond to modernity's disenchantment together with its own dilemmas and difficulties are limited. As Rundell notes '[m]odernity cannot meet its own challenges because it lacks the depth to do so'[22] – we can only draw from our closed world structures. He notes that for Taylor moder-nity's fate is to 'produce a wonderfully monstrous paradox – once the world was discovered to be round and its motion circular around the sun, it became flat and linear'.[23]

In an enchanted world, however, our environmental relationships might provide the means to empower us to explore and come in contact with the meaning in the world. These meanings invite conversations concerning how we relate to others and with the natural environment. Moreover, we are engaged in the natural environment and the various approaches discussed above impact on the way we are anchored in reality. These evaluative and interpretivist stories are built on Taylor's meta-physics and evaluative framework and combine many ideas from different sources, which in turn reflect his approach to how our language builds up richer pictures of humanity's place in the natural environment.

In effect, Taylor introduces us to a conversation that has no ultimate end because it presents a challenge to a type of exclusive humanism – too readily seduced by closed world thinking – that (wrongly) tries to prove that experiment, procedure and testing can be unproblematically applied to human affairs and environmental issues. In other words, the conversation about our place in nature is not legitimately foreclosed by invoking inappropriate procedures or tests and so remains open-ended.

5.4   Taylor, Merleau-Ponty and language

Before we delve further into the connections among interpretation, language and perception it is important to recall that Taylor focuses on the conditions of having what we can call experience. He explores the Heideggerian formulation which points us toward another facet of the same phenomenon, the fact that anything can *appear* or come to light at all. Here Taylor focuses on the role of language whereby our processes of articulation reveal the value of different visions of the world. That is, the expressive tradition articulates the centrality of language as the key to revealing what we perceive and express from the world.

Taylor continues that Heidegger, with his concern to get beyond subjectivistic formulations, ends up talking about the 'clearing' (*Lichtung*) which involves exploring the fact that anything can *appear* or come to light at all. He continues:

> This requires that there be a being *to* whom it appears, *for* whom it is an object; it requires a knower, in some sense. But the *Lichtung* formulation focuses us on the fact (which we are meant to come to perceive as astonishing) that the knower-known complex *is* at all, rather than taking the knower for granted as 'subject' and examining what makes it possible to have any knowledge or experience of a world.[24]

According to Taylor's expressive perspective, language not only designates meaning, but it is also a medium through which conscious thoughts and our perceptions become part of our everyday existence. As outlined earlier in this chapter, our conscious thoughts are parasitic on our basic perceptual abilities that cue us in the natural environment. Further Taylor is extending the interpretivist work, when he explores the objectivity of value in the natural environment. This exploration is undertaken using two interlocking assumptions. First, there is a challenge to the purely designativist view that words are the only vehicle of our conscious interactions and representations of the natural environment. Second, there is the expressive view that there is more to language than designativists assume – that is, language operates against a background context of meaning. These two points reflect how consciousness, or existence, is realised against the backdrop of its location in a body situated in environmental and social contexts. It is our language that makes this possible. Hence language must be probed from an entirely different point of view. Taylor continues:

It is not just a set of signs which have meaning in virtue of referring to something, it is the necessary vehicle of a certain form of consciousness, which is characteristically human, the distinct grasp of things which Herder calls 'reflection' (*Besonnenheit*). In other terms, words do not just refer, they are also precipitates of an activity in which the human form of consciousness comes to be. So they not only describe a world, they also express a mode of consciousness, in the double sense outlined above, that is, they realize it, and they make determinate what mode it is.[25]

The key to Taylor's view is that our being human is not simply consciousness in its most basic form, but is realised against a background context of environmental and social values. In this regard Taylor develops the ideas of Herder and Humboldt to deepen our appreciation of the natural environment.[26] Of course Taylor is aware that Herder began his analysis of language in *Ideas for the Philosophy of History of Humanity* with the claim that 'a people has no idea for which it has no word'.[27] Accordingly, language and words are not simply located in the mind's storeroom, but are created and transformed by people.

Taylor explains that for Herder the existence of this representational or linguistic consciousness is the central issue. It is language which makes it possible for us to have this distinct, focussed awareness of things, where animals remain caught in the dream-like, melodic flow of experience. He emphasises the ways by which our self-interpretations are expressed through the clearing that language creates.[28] Involved in this clearing are the different aspects of reality and the meanings that emerge from the world. The awe and beauty in the natural environment arise in such a vision. Our experience of nature's awe and beauty as beings in the world suggests a re-enchantment of things for us. And so we find Taylor emphasising 'the non-arbitrary, non-projective character of certain demands on us, which are firmly anchored in our being-in-the-world'.[29] These demands do not just emanate from us, but nor are they inscribed in the universe (or the Universe and God) independently of us. They arise in our world.[30] This emphasis on the role of language given by

Taylor is designed to explore the fact that we are self-interpreting beings. Here Taylor returns to his notion of what are strong evaluations and self-interpretations. That is, his alternative social science approach to research involves an account of the person or self that he describes as 'expressivist' or 'self-interpreting'.[31] Moreover, the art of self-interpretation involves challenging the individualism on which modern social science is based and calls for a totally different way to engage with our economic and social realities.

Taylor emphasises four key features associated with the notion of what he calls 'strong evaluation'. These factors impact on how we re-enchant our understanding of the natural environment. In summary form they are:

(a) That such an evaluative response genuinely motivates us – it is not simply a cover or rationalisation, or a screen for some other drive;
(b) It can fail to occur on some occasions or in some people, but this betokens some limitation, blindness, or insensitivity on their part;
(c) In other words, there is something objectively right about this response;
(d) We can and ought to set ourselves to cultivate this response, refine/improve our perception of its proper object.[32]

Taylor then emphasises that it is our ability to provide self-interpretations of our reality that differentiate us from other animals. And such self-interpretations cannot but instantiate distinctions of worth. For Taylor self-interpretations are associated with the notion of strongly valued goods which are something different from weak-order goods which people hold onto irrespective of other more desirable goods (such as wanting ice cream). The values expressed through practical reasoning involve our more significant purposes and cannot be reduced to basic wants and needs. Here, Taylor's Aristotelian sense of practical reasoning involves expressing these values as they shape and influence who we are, and how we make judgements.

In this way Taylor's multifaceted engagement with the transcendental deduction of experience is not limited by our abilities in perception because it is in turn shaped by our very embodiment in the world. Through engaged agency Taylor's metaphysics is implicit in his analysis of secularism and can then be used to respond to Bryan Norton's environmental pragmatism. Norton called for 'a truly inter-disciplinary replacement of our outmoded world-views with new 'metaphysical and epistemological terms'.[33] Interpretivism involves an account of the person focusing on self-interpretations which are not something that can be understood with absolute and foundational certainty.[34] This understanding of interpretivism emerges in our interaction and utilisation of the concept itself as I discussed in Chapter 3. Here Taylor offers an expressive view of language in challenging the dominant analytic supposition that language simply refers words to ideas.[35] On the view that Taylor believes to be correct there is more meaning to be discovered in the natural environment than can be revealed while using a designative approach to language. That is, in language things of intrinsic value and their meaning show up in a context thereby reflecting on humanity's ability to express and utilise ideas in language.[36]

In sum, this section has focused on Taylor's work on language which shares with Jim Cheney the environmental view that: 'the stories we tell are not finished; and they are not our stories only, not simply the constructions of Cartesian and Humean subjectivies, but stories told by the world as well'.[37] For both Cheney and Taylor this fusion of ideas through language reflects various connections between self and the natural environment. This gives rise to an unproblematic realism where environmental awareness emerges as a fruit of democratic, dialogic and interpretivist analysis. More fundamentally, I have argued that it is via perception and through language that the various meanings of the natural environment emerge.

5.5 Conclusions: Natural connections among perception, language and strong evaluations

In sum, the previous three sections of this chapter developed connections among our self-interpretations as they emerge from our perceptual awareness of the world. It was stressed that some interpretations reveal different features of our reality as more important than others which is why for Taylor, interpretivism involves an evaluative framework. This chapter has argued that Taylor's work on our skills in perception put people directly in the natural environment. I used Taylor's work on Merleau-Ponty to explain that our more basic involvement in the world provides a strategy to overcome the strict division between our minds and the world.

The first step involves the idea that humanity has a world, and it is in this world that our perceptions are activated. The second step involves a more basic feature of our humanity and that is the fact that we are not only conscious beings but also beings who have perceptual abilities to engage with the world. These abilities give us access to the world and are the very foundation of having a world.[38] At the heart of this assertion is a point relevant for environmental ethics that humans are body-subjects in natural contexts. Furthermore, in combination Taylor's interpretivism reflects his ontological account of reality which challenges the designative and procedural approaches that lead to closed world thinking. The designative approach offers an ontological description of the natural environment which conceives of it in a mechanistic and procedural way only. As such, the designative ontology of language conceives of the natural environment as merely a means to an end. We have the potential to lose sight of the other meanings that emerge in our interactions with it. Fundamentally, mechanist and procedural approaches glide over the various meanings and values that can emanate from the natural environment.

Part II

6
Taylor's Environmentalism and Critique of Utilitarianism and Instrumental Reason

This chapter outlines Taylor's critique of utilitarianism in an environmental context by returning to some important ideas developed independently by Robert Goodin, Lawrence Johnson, Noah Lemos and Peter Singer. I also examine Taylor's critique of utilitarianism and animal liberation. The key aim is to set out the issues that are needed if we are to improve our understanding of the natural environment.

6.1 Taylor's environmental contribution

Like postmodernists, Taylor is critical of the Enlightenment view of the natural environment, but he comes to very different conclusions about our relationships with it.[1] Taylor helps us unite considerations of the social and the ecological in a manner more fruitful than the theorists who propose adopting utilitarian approaches to environmental ethics.[2] This chapter examines Taylor's criticisms of utilitarianism by focusing on Peter Singer's work on animal liberation and returning to Lawrence Johnson's morally deep world. There is also a growing body of literature that attempts to reinvigorate utilitarianism using Derek Parfit's famous work on population and how to choose which world is the best one to live in.[3]

Like many of these theorists, he is critical of the Enlightenment's instrumental reason and, more particularly, the focus on a procedural *modus vivendi* which characterises Enlightenment-inspired liberalism and utilitarianism.[4] According to Taylor, the Enlightenment emphasised a *punctual* understanding of the human subject,[5] defined in abstraction from any constitutive concerns – in isolation from its community and significant attachments. It endorsed disengagement, seeing people as free if they were independent from external interference. It looked inward rather than to external things – to the *polis* or republic and even less the environment. Morality, therefore, was seen as either self-interest or a de-contextualised 'right'. In eighteenth-century terms, which many philosophers have now forgotten, the Enlightenment collapsed ethics into morality.[6] Morality (*moralis*) – being true to a role – was personal, whereas ethics was about objective value structures (as in Hegel's *Sittlichkeit*). Enlightenment thinkers such as Immanuel Kant reduced ethics (*ethicus*), where the focus was on *what I should be*, to morality where the focus was *what I should do*. Kant's categorical imperative acted as a yardstick for what *I ought to do* rather than what it was *good to be*.[7] Considerations of good structures (which might be environmentally embedded) were displaced by considerations of principles of right behaviour.[8]

The Enlightenment celebrated autonomy rather than authenticity, which affirms self-determination but not in isolation. Autonomy centred on a negative conception of freedom based on an atomistic conception of human nature.[9] People, therefore, were free when they stood apart from nature. That separation could easily turn into an affirmation of domination of both the natural environment and other people. Alisdair McIntyre criticises the fact that the modern citizen 'is a citizen of nowhere, an internal exile wherever he lives...modern liberal society can appear only as a collection of citizens of nowhere who have banded together for their common protection'.[10] Rootless associations are hardly likely to be ecologically aware.

Interpretivists, Taylor feels, must back-track a little, consider how people reflect on environmental issues and appreciate the extent to which nature may be seen as a source of the self,[11] submerged by atomistic, pre-social and ahistorical conceptions of humanity. Taylor's aim is to show that nature is not just a constraint on freedom but helps to structure freedom. In this chapter I return to Taylor's work on the old argument that classical liberalism and utilitarianism connect with free-market arguments. These arguments assume that, through incentive structures, humanity can rectify ecological damage. An interpretivist-designed and inspired environmental ethic, on the other hand, must go further in charting the way that collective institutions may show how an appreciation of nature is necessary for the attainment of the good life. To this point I have argued that an interpretivist approach can embrace the idea of nature's intrinsic value, and this chapter examines key political issues emanating from utilitarian quarters.

Following this introduction is Section 6.2 which outlines Taylor's critique of utilitarianism to introduce his work on practical reason and the natural environment. Section 6.3 explores whether higher-order goods exist using some ideas from Derek Parfit and Noam Lemos. Section 6.4 outlines Taylor's critique of utilitarianism and animal liberation. In Section 6.5 I review Taylor's key environmental visions. In Section 6.6 I conclude the chapter with some extensions to interpretivism to incorporate the insights of other political orientations be they Post-Hegelians, Earth Firsters and counter-culturalists who advocate the preservation of the diversity of life on earth.

6.2   Sources of value in nature: Taylor's critique of utilitarianism

It is necessary to explore utilitarianism because it has been used to justify a number of influential environmental and animal rights perspectives.[12] For example, the utilitarian thinkers that were inspired by the Enlightenment of the eighteenth century likened the processes of nature to the workings of a machine. They assumed that humanity had the technological capacity to solve the problems confronting it. Human beings were considered to be at the centre of the world which was explained by Collingwood in his nineteenth-century Hegel-inspired *Idea of Nature*, which held that mind makes nature; 'nature is…a by-product of the autonomous and self-existing activity of mind'.[13] Instrumentalism was also evident in John Stuart Mill's 1854 essay 'Nature'.[14] For Collingwood and Mill, the maxim to 'follow nature' sounded like a return to an animistic ethic. Mill argued that it is ambiguous because humanity has the capacity to master it:

> Everybody professes to approve and admire many great triumphs of Art over Nature: the junction by bridges of shores which Nature had made separate, the draining of Nature's marshes, the excavation of her wells, the dragging to light of what she has buried at immense depths in the earth; the turning away of her thunderbolts by lightning rods, of her inundations by embankments, of her ocean by breakwaters. But to commend these and similar feats, is to acknowledge that the ways of Nature are to be conquered, not obeyed: that her powers are often towards men in the position of enemies, from whom he must wrest, by force and ingenuity, what little he can do for his own use, and deserves to be applauded when that little is rather more than

might be expected from his physical weakness in comparison to those gigantic powers.[15]

According to Mill nature is not only a barrier but is humanity's antagonist.[16] Though modern environmental theorists are more aware of the dangers of instrumentalism and anthropocentrism, but one can still find echoes of Mill's views among modern 'shallow green' theorists who are sceptical of radical and Deep-green Ecology.[17] It is easier to talk about the interests of nonhumans; and the question of interests which is the stock in trade of utilitarian theorists. Indeed, Peter Singer, one of the most famous exponents of animal liberation, uses utilitarian reasoning[18] and so does Robert Goodin, who is prominent among philosophers who have studied the environment.[19] It is very difficult to consider the rights of nonhumans when these rights clash with policies designed to alleviate inequality and suffering in human communities. Nevertheless, a few notable works adopt a rights-based approach to animal liberation[20] and Deep ecologists sometimes talk about the rights of ecosystems.[21] Lawrence Johnson has argued that if ecosystem interests can be identified then it becomes possible to extend rights to them. In effect, Johnson's utilitarianism is used to make an argument for the extension of rights to environmental ethics. Moreover, it remains problematic whether ecosystems have rights in the classical liberal sense, because rights are a human-centred construct as language theorists have pointed out.[22]

Lawrence Johnson's *A Morally Deep World,* for example, offers a philosophical investigation into the moral worth of ecosystems. Living systems, he argues, are entities in a persistent state of low entropy. They are sustained by metabolic processes which accumulate energy. Their organic unity and self-identity are maintained in equilibrium by homeostatic feedback systems.[23] Johnson essentially offers utilitarian means to rank interests. Starting from the view that 'what we value...expresses what we are', he extends his purview to the moral consideration of other entities.[24] It is in those depths that our well-being interests have their roots. Even among humans, neither rationality nor sentience is a necessary condition for the moral significance of an interest. For us to recognise the moral standing of only those beings who are like us in being human, rational, or sentient would be arbitrary and morally unjustifiable.[25] In effect, Johnson's utilitarianism is used to make an argument for the extension of rights to environmental ethics.

Furthermore, if John Stuart Mill's essay 'Nature' deprecated the notion of 'following nature' and advocated humans dominating it, humans also possessed the ability improve it.[26] Here Robert Goodin agrees with the irrelevance of the maxim 'follow nature', but he is less sanguine than Mill about the consequences.[27] Seeing nature's intrinsic role, Goodin notes that value in nature is created by processes which are larger than, and thereby necessarily beyond, humanity's comprehension. Human intervention may detract from nature's value with the effect that an object loses 'that value-imparting property once it has been restored, replicated or recreated through human interventions'. It is no longer authentic.[28]

Goodin's argument is an interesting departure from a classical liberal view on the natural environment. Yet, utilitarian reasoning limits practical reasoning to instrumental considerations and has had little to say about 'intrinsic value', much less authenticity. That is not the case with Goodin. His view, however, gives rise to an instrumental conception of authenticity where it is important that we assign weights to the different options that confront us. Nature, according to Goodin, has intrinsic value which affects the weight humans apply to it. Humans are constantly balancing their own interests against nature's interests. Thus, we are presented with a system of trade-offs where nature's gain is humanity's loss, and vice-versa.[29] It is a zero-sum picture in which public policy advocates cannot pick or choose between parts of the theory.[30] His opinion, moreover, that utilitarianism is simply a valid public policy which does not correlate with personal morality,[31] leads to the view that, one can affirm the value of nature while not having to adopt a different lifestyle. He advocates:

(a) pursuing green public policies, to secure the larger natural contexts in which we want to see our own projects set; but,
(b) refusing to adopt green personal lifestyle recommendations aiming to deprive us of the distinctively personal stance from which harmony with nature would be satisfying or even meaningful.[32]

We are left with little analysis of the structural and social causes of environmental problems and an endorsement of current liberal-democratic institutions. Goodin's cost-benefit analysis, therefore, tends to absolve citizens from personal responsibility for environmental problems.

Goodin's approach, which remains within the framework of instrumental reason, is clearly different from that of Taylor who is an advocate of new social relations and is also extremely critical of utilitarianism. Consider Taylor's discussion of authenticity that was defined earlier as autonomy within context. But the context is not an instrumental one.[33] Authenticity is not something which may be weighed or traded.

It is about people being-in-the-world and is something different from economic notions of choosing how to maximise the options which confront us. Of course it is important to remember that it is very difficult to consider the rights of nonhumans and ecosystems when these rights clash with policies designed to alleviate inequality and suffering in human communities.[34] Nevertheless, a few notable works adopt a rights-based approach to animal liberation[35] and Deep ecologists sometimes talk about the rights of ecosystems.[36] For example, I have already mentioned how Lawrence Johnson developed an argument that if ecosystem interests can be identified then it becomes possible to extend rights to them.

As stated, ecosystems do not have rights in the classical liberal sense and this is because they are a human-centred construct.[37] Moreover, authenticity is not something which may be weighed or traded. It is about people being-in-the-world and is something different from economic notions of choosing how to maximise the options which confront us. Long ago Mark Sagoff put it well:

These values we cherish as citizens express not just what we want collectively but what we think we are: We use them to reveal to ourselves and to others what we stand for and how we perceive ourselves as a nation. These values are not merely chosen; rather they constitute and identify we who choose; Wilderness, rivers, estuaries, bays, forests, and farms have voices. They express our shared values and transmit them. They speak for us.[38]

Sagoff identifies the intrinsic value which many environmentalists find appealing. Though it is problematic whether nature possesses such intrinsic value independent of a valuing subject, it is equally questionable whether nature can be valued using simple utilitarian criteria.[39] Unrestrained utilitarianism appeals to the lowest common denominator in humanity, losing sight of what humanity has come to value in nature. Taylor's critique of utilitarianism is trenchant:

But when one moves from opposition of government, from attack to building a new order, then it has repeatedly been manifest how thin and at the same time how threatening the utilitarian outlook is. Thin, because building requires some sense of the goods one is for and not only what one is against. Threatening, because the refusal to define any goods other than the official one of instrumental efficacy in the search for happiness can lead to appalling destruction in a society's way of life, a levelling and suppression of everything which doesn't fit in that tunnel vision, of which the modern consequences of bureaucratic rationality offer ample testimony, all the way from the Poor Law of 1834 to the catastrophe of Chernobyl.[40]

Utilitarianism gives undue emphasis to appetite where the good life is one reflected in continuing escalations in living standards. According to utilitarian reasoning, moral worth is determined by appeal to the appetite of unregenerate individuals. Taylor reminds us that Plato knew this lesson well.[41] The utilitarian appetite tends to run to infinity unless controlled by reason. Consumer society, according to Taylor's interpretivism, appeals to the lowest common denominator in us where nature is seen as merely a 'standing reserve' to satisfy individual basic wants and needs.[42] For Taylor, this awareness cannot be captured in an instrumental model such as utilitarianism. Taylor quotes Mill's *On Utilitarianism* that for Mill:

'Questions of ultimate ends are not amenable to direct proof', he avers, and yet Mill says 'considerations may be presented capable of determining the intellect either to give or withhold its assent to the doctrine [namely, of utility]; and this is the equivalent of proof.'[43] This may sound like someone trying to squirm his way out of a contradiction, but the distinction is quite clear and sound. You cannot argue someone into accepting an ultimate end, utility or any other, if he really rejects it. But, in fact, the case of utilitarians is that people do *not* reject it, they all really operate by it, albeit in a confused and therefore self-defeating fashion.[44]

Mill points out that the question of ultimate ends is open-ended and cannot be proven to anyone. Further, Mill argues that in fact people do not reject this objective, and this is the equivalent of proof because people desire happiness. This utilitarian doctrine has become ingrained in modern Western societies together with the idea that rights are fundamental and inviolable. This process, I argue, creates an inward culture which leads us to value nature only in its absence.

For Taylor, the alternative is modern liberalism's reliance on *apodeictic* forms of practical reasoning that assumes our actions can be framed as maxims and then adjudicated on the basis of universal criteria independent of culture and society. Ever since Moore's refutation of Mill's so-called proof of utilitarianism, the alternate *ad hominem* mode of reasoning has been defined primarily in terms of weak-form evaluations, such as pleasure and pain. This gives rise to a further problem with utilitarianism in that it cannot judge between weakly valued and strongly valued goods as they frame our choices and influence our beliefs. Thus, strongly valued goods are something different from weak-order goods which people hold onto irrespective of other more desirable goods (such as wanting ice cream). The values expressed through practical reasoning involve our deeper purposes and cannot be reduced to basic wants and needs. Here, Taylor's Aristotelian-inspired sense of practical reasoning involves articulating and expressing these values as they shape and influence who we are, and how we make judgements.

Taylor explains that the contradictions in Mill's model reflect a utilitarian bias in modern society. He suggests that Mill's circular argument is reliant on a weak-form *ad hominem* justification that appeals to a person's prejudices, or special interests instead of to a person's intellect.[45] In Taylor's view Mill's refined version of liberalism, therefore, does not come to grips with the stronger-order goods that frame and impact on our identity. These are the goods of significance which reflect culture, environment and family relationships; more particularly, these values cannot be reduced to economic calculations. Utilitarian reasoning, which could easily be used to support capital accumulation, has the potential to submerge other values. It can 'just degenerate into organised egoism, a capitulation before the demands of our lower nature'.[46] As noted in Chapter 1, Taylor's critics claim that his view of utilitarianism is over simplified and out of date.[47] His answer is:

> But such 'hard line' utilitarians aside, I agree that most of the proceduralists I am attacking have a place for notions of the good life, and many are moral realists like myself. My grievance against them is that they foreshorten the scope of moral philosophy, pay risibly little attention to the good, and concentrate largely on the principles by which we can determine the right.[48]

Central to the development of an interpretivist-inspired environmental ethic is the contention that of all of the procedural principles the most suspect is that which considers authenticity to be the maximisation of consumption and pleasure.

6.3   Higher-order goods and nature

In that context, consider Parfit's dilemma.[49] Parfit asks us to consider choosing between life in a drab eternity or a century of ecstasy. By choosing the drab eternity we choose a life of muzak and potatoes in which there are no 'higher-order' goods to consume or even contemplate.[50] By choosing the century of ecstasy we choose the best life available but also to accelerate the rate at which we use the earth's resources. The obvious choice is the century of ecstasy. '[T]hough each day of the Drab Eternity would have some value for me, *no* amount of this value would be as good for me as the Century of Ecstasy.'[51] He rejects:

> The Repugnant Conclusion: Compared with the existence of very many people – say, ten billion – all of whom have a very high quality of life, there must be some much larger number of people whose existence, if other things are equal, would be better, even though these people would have lives that are barely worth living.[52]

Apparently some goods are better than others and yet choosing the century of ecstasy raises huge ethical questions. Lemos's arguments are useful here because he asks us to consider a world that contains a million people living pleasant and morally virtuous lives, pursuing aesthetically and intellectually excellent activities.[53] Then, we are able to consider a different possible world called W'. The sole sentient beings are worm-like creatures which Lemos names O-worms. The O-worms have no self-consciousness; no awareness of the past or future; no friendship or love for one another; and, of course, no moral, aesthetic, or intellectual life whatsoever. They never feel any pain, but they do feel intense pleasure for a few seconds on the sole occasion in their life cycle when they reproduce. W' is a dull world, but it does seem to be a good one, containing a vast number of instances of intense pleasure. Lemos concludes:

> When one reflects on these worlds, it is plausible to think that no matter how many instances of pleasure W' contains, it is not intrinsically better than W. Speaking roughly, if there were another world just like W', but containing more O-worms and more instances of pleasure, it would be better than W', but it would be no better than W. If W is intrinsically better than W' no matter how many instances of pleasures we find in W', then it does seem that Bretano, Ross and Parfit are right about the existence of a high class of goods. It might be hard to identify exactly what belongs in this higher class, but if W is better than W' we can reasonably think that there are such.[54]

If higher-order goods exist, political theory should do away with the utilitarian idea of trying to measure the worth of different states in terms of happiness.[55] But should one measure them in terms of perfection?

Or is the notion of measurement itself otiose? And what are the criteria for perfection? In this way, communitarian inspired interpretivism can only offer considerations about how people can lead better lives. These must be historically determined and arrived at by hermeneutic procedures and democratic deliberation. The demos may get them wrong but there is no need to assume, as some pre-democratic writers insist, that a concern for appetite means that it will always get them wrong.

What then are we to make of the injunction to 'follow nature'? Twentieth-century analytical philosophers might consign the notion that nature possesses value to the category of *ought* statements concerning what people should do rather than a statement of fact. Others might say that simply to follow nature is hardly realistic in the modern world. But even if it is unrealistic, one does not have to jump to the conclusion that it is always better to impose *humanity's imprint* on nature.[56] That was the position of Aldo Leopold's famous 'land ethic'[57] which has been celebrated by many Deep ecologists. Leopold insisted that '[a] thing is right when it tends to preserve the integrity, stability, and beauty of the biotic community. It is wrong when it tends otherwise.'[58] Was this a radical rejection of Enlightenment anthropocentrism, as J. Baird Callicott and other Deep ecologists maintain, or was it just a warning against its instrumentalism?[59] On that question ecologists are divided.

Taylor evokes memories of Rousseau, noting that humanity's '[c]onscience is the voice of nature as it emerges in a being who has entered society and is endowed with language and hence reason'.[60] Surely a sense of intrinsic value occurs when people confront the natural environment. That sense, moreover, is unique, particular and historically located[61] – a fact not easily addressed by many hard and soft ecologists unfamiliar with the romantic counter-Enlightenment ideas associated with the work of Hamann, Herder and Humboldt.[62] It is easy to see why that is the case. John Rawls notwithstanding,[63] it is unfashionable nowadays to attempt to construct a philosophical position out of an intuitive sense. But that approach was once common. Many eighteenth-century thinkers were concerned with showing that humans were endowed with a moral sense, 'an intuitive feeling of right and wrong'. In the words of Taylor:

> The original point of the doctrine was to combat a rival view, that knowing right and wrong was a matter of calculating consequences, in particular those concerned with divine reward and punishment. The notion was that understanding right and wrong was not a matter of dry calculation, but was anchored in our feelings. Morality has, in a sense, a voice within.[64]

Nor is romanticism all that fashionable. But romantic rebellion continually recurs. As discussed in Chapter 3 this dilemma involved two powerful aspirations – to expressive unity and to radical autonomy. It was argued that these dilemmas have remained central to preoccupations of modern environmental and social politics. While it may not be possible to combine them it is important to understand how they have been integral to Marxism, Anarchism, technological Utopianism or the return to nature.[65]

To affirm morality's voice within us and to sympathise with romanticism, however, is not the same as affirming nature's intrinsic value independent of human beings – a step Taylor is unwilling to take.[66] For him intrinsic value is a human-centred concept which reflects the values of the language bearer for whom nature possesses such value. Accordingly, it is useful to approach the value of the natural environment through practical reasoning about freedom and how nature structures the possibilities of freedom, charts democratic-republican processes which may recover them, and shows self-interpreting humans other ways of being-in-the-world.[67]

As stated in earlier chapters, Taylor never presents a full philosophical defence of intrinsic value but develops connections and interpretations through language. Therefore, through his critique and metaphysics, he does offer us ways of thinking about the relationship between societies and a sense of value. Of importance here is the idea of insight into the values of different communities, which might highlight elements of a common appreciation of intrinsic value. Moreover, he is concerned with what he calls 'strong evaluations' such as cultural and environmental relationships which shape people's identities and which cannot be captured by philosophical perspectives which operate within closed world thinking as outlined earlier in Chapters 2 and 3. By recognising the importance of ethical and evaluative frameworks, he shows the dialogic frameworks that may realign humanity's understanding of what nature means and, through modern communicative structures, may take some steps towards moderating its impact on nature.

6.4 Taylor's critique of utilitarianism and comments on animal rights

Having sketched out the similarities and differences between Taylor and ecologists, we should consider the extent to which his ideas converge with those of advocates of animal liberation. Like them, Taylor lays considerable emphasis on sentience. Sentience reflects what Singer referred to as the capacity to feel pleasure and pain; the capacity to enjoy life; where 'enjoying' is used in its old-fashioned or legal sense, as in 'enjoying the use of one's limbs', rather than in its narrower colloquial sense of having a good time. Taylor's interpretivism is critical of this perspective and argues that sentience involves self-awareness of self-feeling. Thus, the intuition underlying a prohibition on killing animals is that this capacity ought to be respected wherever it exists and that one ought not to snuff it out or seriously impair it in beings who have it.[68] Nonhuman animals, as sentient beings, are to be respected but do they have rights? Taylor explains:

> Why ascribe rights to animals? Or if this sounds too bizarre, why claim that it is wrong to kill or to inflict pain on animals? The answer commonly given is that they are sentient beings. But this concept of sentience is not as simple as it looks, as those who have argued this position readily acknowledge. We cannot take it to mean simply 'capable of feeling pain', and argue from that common factor that we ought not to inflict pain on any sentient being; because in fact we want to claim more: we would not agree, if there were some utterly painless way of killing, say by way of a laser-ray, that vegetarians ought to drop their objections to killing cattle for food, let alone that it would be licit to kill people for the convenience of others or for the demands of progress (though in the human case, the argument would be complicated by the anxiety about being a potential victim of the laser-ray).[69]

Sentience, itself, is not a necessary nor is it a sufficient criterion for the development of a rights culture. In contrast to utilitarianism, Taylor endeavours to determine in Hegelian fashion what criteria might be useful. According to interpretivism one criterion is that of authenticity which can be worked out through the capacity to command respect. In this way Taylor is moving beyond strict liberal conceptions of a rights culture. For Taylor, the liberal injunction that rights are sacrosanct can be challenged by considering the conviction that certain beings have rights; and, this is a reflection of their capacity to command and give respect.

Therefore, rights presuppose the capacity for reciprocity itself, a capacity unique to language-bearing entities who choose and determine the shape of the rights. Taylor extends the notion of rights and argues that this supposition involves the capacity to command respect. Taylor continues:

[T]hen indeed, it is sufficient that we identify A as possessing this capacity to make A a bearer of rights'. It is clearly always a necessary condition as well of bearing those rights, if only because what the rights are to will be defined in relation to the capacity. But the mere possession of the capacity will have no normative consequence at all for us if we do not share the conviction that this capacity commands respect. [70]

According to Taylor, therefore, the capacity to command respect is clearly always a necessary condition for rights.[71] Nature may deserve respect, but can it *command* it? Surely respect seen in terms of a claim implies both recognition and reciprocity.[72] Rights, therefore, belong to the realm of communicative action – a conclusion one would expect from someone who lays considerable stress on the importance of language and practical reason.

Here, Taylor's sense of practical reasoning and his work on all that is significant for a full life involves exploring the role that the natural environment plays in the development of human freedoms. Taylor offers us the criterion of 'significance' in shaping human identity.[73] The strength of people's motivations depends on the significance that they assign to various decisions. In evaluating motivation, there are 'weak' and 'strong' approaches which emerged from the Enlightenment and all the changes it brought about. Taylor argues that:

The Enlightenment gave rise to a new kind of indignant protest against the injustices of the world. Having demolished the older visions of cosmic order and exposed them as at best illusion, and perhaps even sham, it left all the differentiations of the old society, all its special burdens and disciplines, without justification. It is one thing to bear one's lot as a peasant if it is one's appointed place in the hierarchy of things as ordered by God and nature. But if the very idea of society as the embodiment of such a cosmic order is swept aside, if society is rather the common instrument of men who must live under the same political roof to pursue happiness, then the burdens and deprivations of this station are a savage imposition, against reason and justice, maintained only by knavery and lies.[74]

Because of its atomistic ontology, liberalism operates with a 'weak' explanation of the motivations behind decisions.[75] Indeed, liberals conceive of decision-making structures which ignore motivations and are intended to be neutral, but it is these motivations which invariably impact on others and the natural environment. Interpretivists, on the other hand, seek a more comprehensive grasp of the assumptions, ideas and values as they frame different cultural, environmental and social frameworks. The understanding gained, Taylor insists, does not necessarily have to lead to specific advocacy. Significance may or may not result in action. Nor need it contribute to a monolithic view of the common good, fixed for all time – a view, one might add, mistakenly attributed by some to Aristotle. Significant elements may form part of a potentially mean-ingful totality which contain *lacunae* or missing pieces of evidence to be reconstructed through historical inquiry.[76] Those significant elements may be fluid, but they are not purely subjective to be approached in a relativistic manner. They may be ascertained, moreover, in hermeneutic fashion – in the same manner as text.[77]

Taylor's hermeneutic notion of understanding here is the fact that elements of significance will unavoidably clash. Obviously one of the greatest contemporary dilemmas is the clash between autonomy and 'expressive unity' which is glaringly apparent in framing the signifi-cance many people attach to nature. A constantly recurring theme in modernity and its political discourse has been the attempt to combine autonomy and expressive unity and the inevitable tensions which arise. Hegel, Taylor reminds us, offered the basis of such a synthesis through the development of reason and spirit in history. In the end, however, Hegel probably failed and few people nowadays would seek a reconcili-ation through Spirit (*Geist*). But the attempt at reconciliation in general is worth making and, at least, considerations of authenticity might keep alive the possibility of reconciliation rather than resigning ourselves simply to instrumental rationality where nature is considered only as the object of consumption – standing in wait for humanity.[78]

6.5   Taylor's directions for environmental politics

At the very least, Taylor's Hegelian approach makes us aware of the diffi-culties involved in reconciling human liberty with nature's intrinsic value.[79] Reconciling the clash between expressive unity and autonomy is ultimately much the same as reconciling different cultural traditions. Hence his interest in language as a means to disclose the otherness of being and the values of significance that have been submerged within modernity's preoccupation with procedure and technology. Perhaps a total reconciliation between humanity and the natural environment is impossible, but one should not resign oneself, in extreme postmodern fashion, to the idea that we all speak ultimately untranslatable languages. Earlier sections of this chapter were motivated by C. P. Snow's 'two cultures'[80] which posed a seemingly unbridgeable gap between science and humanistic discourses:

> [C]ertainly the rational science model dissociating understanding and attunement has wreaked havoc in its successive misapplications in the sciences of man in the last few centuries. But this again says nothing about its validity as an approach to inanimate nature.[81]

That gap is now closing to some extent. Here it is worth reflecting on the Heideggerian insights that Dreyfus and Taylor developed. Independently, both Dreyfus and Taylor argued how important are the background structures of our value systems which shape the humanities and natural sciences.[82] Taylor, in particular, argued that the background practices and structures of the social scientist cannot be taken for granted and ignored in precisely the way the background was ignored in natural science.[83] Indeed, Taylor's interpretivism is committed to the perspective that both the natural and social sciences need to consider their traditions and practices. Taylor's argument, therefore, involves modifying Snow's suppositions and creating a bridge between science and the humanities. Taylor argues that the limitations of frameworks which ignore the background thereby offer only a one-sided and narrow appreciation of humanity, which was my argument in Chapter 2.

These themes converge with his work on value in the world and the notion that 'there is no such thing as a fixed horizon' for political systems to construct. For Taylor:

> Not that it's all going to break down, but that capitalism creates pushes towards ecological ruin, inequality, higher Gini coefficients. The world intelligently understood is a world humanely understood; that we try to manage these tremendous tensions and get as many of these threatened values saved while not killing the goose that's laying the somewhat half-done eggs.[84]

Implicit in Taylor's work is an exploration into the structures of governance that would assist the understanding of the role that the natural environment plays in the development of human freedoms, Taylor offers us the criterion of significance in shaping human identity.[85] The strength of people's motivations depends on the significance they assign to various decisions which include for our purposes the natural environment. It is worth revisiting Taylor's work on evaluating motivation as 'weak' and 'strong' approaches to reasoning, which was discussed in Chapter 3. I pointed out that strong evaluations are not just a 'condition of articulacy about preferences, but also about the quality of life, the kind of beings we are or want to be. It is in this sense deeper'.[86]

6.6   Conclusion

The above discussion examined Taylor's criticisms of the connections between utilitarianism and liberal instrumentalism. I focussed on the type of reasoning used to arrive at ecological consequences, noting that within its *modus operandi*, procedural approaches such as liberalism and utilitarianism have a tendency to displace investigation into more substantive relationships between humanity and nature. The discussion examined the importance of not only Taylor's metaphysics but also his political work for such an investigation because no longer can nature be viewed as a malleable input into production. Factually, far-from-equilibrium systems theorists and many others tell us, it is doubtful whether humanity has the technological capacity to control and master nature's processes. Even if it has, a consideration of nature's intrinsic value is important. This chapter has pointed out that Taylor's political work leads to an awareness of nature's 'voice within' which should not be dismissed as out of date romanticism.

The question now arises and will be pursued in later chapters as to whether representative democratic systems provide avenues adequate to considerations of ecology and environmental politics. More broadly do they provide avenues for discussion and dissemination of the various sources of modern identities? Interpretivism seems to offer hope on this score.[87] It invites us to think about the means through which citizens might have direct access to the deliberations of governments. It invites us also to think about the dangers of free-market liberalism to communities and communal sources of the self – which must include ecological considerations. Put another way, it squarely addresses questions of identity which must have ecological determinants. The chapters that follow will examine in more detail the insights of other political orientations be they Post-Hegelians, Earth Firsters and counter-culturalists who advocate preservation of the diversity of life on earth and question the fundamental tenets of modernity.[88] The environmental aim is to incorporate other frameworks including insights of communitarians, interpretivists, liberals and utilitarians thereby opening us to the existence of other values.

7
Taylor's Critique of Instrumentalism, Liberalism and Procedure in Politics

This chapter examines Taylor's critique of liberalism and then extends it to the ecological dimension within modern political discourse. It is argued that extending liberalism to encompass relationships with the natural environment does not fully come to grips with the dilemmas confronting humanity.

7.1 Introduction

This chapter focuses on Taylor's criticisms of procedural liberalism on its understanding of the environmental implications of his inter-pretivist approach.[1] Taylor's central critical focus has been on John Rawls' theory of justice.[2] Many environmentalists have extended Rawls work to ecological politics and this involves determining whether Rawlsian political liberalism can create changed attitudes toward the natural environment. Taylor observes that at the height of the Rawls' boom nobody thought Rawls was 'out of his mind to try to decide between different principles of justice with the aid of rational choice theory'.[3]

For Taylor, during the past three decades criticisms of Rawls were in a distinct minority, as the Rawls industry showed. This chapter aims to examine how Taylor's reforms might be integrated into the basic institutions of liberal democracies. The chapter develops the theme that humanity and the natural environment are entwined and cannot be separated in the long term. The argument involves relating humanity with the natural environment without infringing on the rights of all people for a decent standard of living. Rawls' focus on rights is the great strength of his political theory, but the crucial question is whether Rawls responds adequately to the new saliences. In his later work he is clearly aware of the omissions in *A Theory of Justice*:

> ...*Theory* leaves aside for the most part the question of the claims to democracy in the firm and the workplace, as well as that of justice between states (or peoples as I prefer to say); it barely mentions retributive justice and the protection of the environment or the preservation of wildlife. ...The underlying assumption is that a conception of justice worked up by focusing on a few long-standing classical problems should be correct, or at least provide guidelines for addressing further questions.[4]

To deal with these issues Rawls offers a 'constructivist' framework which some theorists have adapted to considerations of nature through an analysis of what Rawls calls 'primary goods'.[5] But is this incorporation of communitarian and interpretivist ideas sufficient to deal with ecological concerns and does Rawls' method conceptualise authenticity? The answers to these questions are to be found in the extent to which Rawls incorporates authenticity under the rubric of what he calls 'respect' and 'self-respect'.

Taylor argues that liberalism tends to homogenise identity because it has been extended beyond the bounds for which it was originally intended. This is an implication of liberal extensionists such as Derek Bell[6], Will Kymlicka[7] and John Meyer[8] who adapt and examine central features of the liberal doctrine to address cultural and environmental political issues. This point has also been noted by Michael Ignatieff in *The Warrior's Honour* when he points out that we are the heirs of a universalising language, 'one that speaks of all human beings enjoying the same rights – that was never intended to include all human beings'.[9] It seems that liberalism has been extended beyond its original bounds tending tacitly to see rights as Jeremy Bentham's 'nonsense on stilts'.

Section 7.2 places Rawls' liberalism in the context of environmental politics. Section 7.3 examines arguments that extend liberalism using the work of Bell, Kymlicka and Meyer. Section 7.4 reviews Taylor's criticisms of procedural liberalism in terms of concepts of respect and self-respect in a collective and environmental context. Section 7.5 considers how Taylor's work can inform new democratic structures to accommodate environmental values. Section 7.6 concludes the chapter.

7.2 Rawls' early work: Liberal-democratic solution to environmental issues

One of the most prominent exponents of liberalism, John Rawls, created a social theory which was intended to reform capitalism by creating fairer relationships between people in a civil society. Rawls attempted to avoid utilitarian criticisms by grounding Kant's categorical imperative within the public culture of a liberal-democratic society. He explained: '[A]s an ideal conception of citizenship for a constitutional democratic regime, because it presents how things might be, taking people as a just and well-ordered society would encourage them to be'.[10]

It is important to remember that Rawls noted that his work engaged ideas inspired by the philosophy of Immanuel Kant. The latter's essential theme was that each individual was sacrosanct and their rights were inviolable. Kant's focus was on the individual and principles of justice. Indeed, Kant had very little to say about animal and environmental rights other than stating that it was wrong to be cruel to animals. He formulated the idea of a 'categorical imperative', where no person was to be treated as a means to an end: 'Act in such a way that you always treat humanity, whether in your own person or in the person of another, never simply as a means, but always at the same time as an end.'[11] In applying a Rawlsian inspired version of Kant, it is argued that the diversity of individual interests must be addressed as a moral term.[12] While Rawls adapted Kant because he believed that justice must be grounded in the public sphere, the overall set of political values remained constrained by principles of political neutrality. The question is whether the environment is a necessary condition of our being and how public spheres adopt environmental awareness. Is the aim to liberalise environmental issues in terms of justice and rights and, in this way, side-step claims that the doctrine is incompatible with environmental issues and local community politics?

However, Rawls' *Theory of Justice* has subsequently been adapted to address environmental issues as well as some of the criticisms outlined in the preceding chapters. Rawls is a trenchant critic of utilitarianism[13], but despite the affirmations of some of his admirers, his starting point is not rights but intuition.[14] Though the criterion of intuition could be an important aspect of any bridge between procedural liberalism and interpretivist authenticity, Rawls himself is a procedural thinker. Rawls began his political journey with a concern for procedure and decision-making in ethics, returning to it in *Political Liberalism*. In *Political Liberalism* Rawls bracketed the metaphysical from the political,[15] offering a number of pragmatic recommendations to reform modern liberal democracies. To deal with these issues Rawls offered a 'constructivist' framework which some have adapted to considerations of nature through discussion of what Rawls called 'primary goods'.[16] According to Rawls, each citizen must ultimately choose and order the goods that are important for their life plans. He developed the concept of a primary good to fulfil each person's personal life plan. Here, Rawls was operating with what might be referred to as an internal reasons approach to political matters but he did not deny the possibility that external reasons might exist and command people's allegiance.

Liberal theorists emphasise individuality, autonomy, and the inviolability of human rights for people to choose their own life plans as they see fit. Moreover, the ideology has an emphasis on the rule of law and in upholding the freedom for people to determine principles of justice using their own sense of individual practical reasoning. Liberals believe that human capacities allow individuals to abstract away from circumstances and contingencies to construct fair and just principles, which they can then use to guide their relationships and interactions with others. Such Rawlsian practical reasoning therefore enables a logical person to correctly and consistently arrive at principles of justice in any given situation. Rawls stated:

> The guiding idea is that the principles of justice for the basic structure of society are the object of [what is known as] the original agreement. They are the principles that free and rational persons concerned to further their own interests would accept in an initial position of equality as defining the fundamental terms of their association. These principles are to regulate all further agreements; they specify the kinds of social co-operation that can be entered into, and the forms of government that can be established. This way of regarding the principles of justice I shall call justice as fairness.[17]

This strategy reflects Rawls' commitment to constitutional principles, and a move away from a fixed model of justice to a more situation-based framework of reasoning. Additionally, the system Rawls focussed on created a system of political neutrality. Despite the many complex moves Rawls made in *A Theory of Justice* to *Justice as Fairness: a Refinement*, the fundamental liberal doctrine remains locked into the current political system. This argument is made noting the urgencies of the environmental crisis.

Long ago, Sibyl Schwarzenbach argued that Rawls had moved in an interpretivist direction.[18] The question is, however, does he bridge the liberal-interpretivist divide? Nevertheless it is important to examine whether liberalism can accommodate such environmental dilemmas. A key assumption of Rawls' liberalism used by environmental thinkers is that justice as fairness is one means to resolve environmental differences. That is, justice as fairness is a decision device that is applicable to and possessed by each individual. Many theorists have worked to extend liberalism to cultural and environmental differences in a society where free citizens are able to live according to their own value systems, and as they see fit. This gives rise to an overlapping consensus which is based on there being a morally significant core of commitments common to the 'reasonable' fragment of each of the main comprehensive doctrines in the community. It remains problematic, however, whether a liberal system committed to an overlapping consensus can create a vision of the world which engages with the notion of significant environmental values and their intrinsic value.

7.3   Extending liberalism to environmental politics

There has been not only a Rawls boom but a proliferation of adaptations of Rawls' key arguments. Of note for present purposes are the various arguments made by Bell, Kymlicka and Meyer who adapt and examine central features of the liberal doctrine. In their various works they extend liberalism to address cultural and environmental political issues. It is also important to remember however that in his later work, Rawls observed the need to alter the principles of justice to accommodate salient issues such as workplace reform, the environment and international relations. These doctrines may include religion, political ideology or morals as I discuss in later sections of this chapter.

At first glance, these arguments are significantly at variance from Rawls' liberalism, in that Rawls advocated the belief that all societies would independently arrive at a set of values and principles that would be viewed as common to all 'reasonable' men and women in different cultures. The need to adapt and extend the liberal framework seemed unnecessary. However, in the context of the environmental theme outlined in this book, it is Bell who outlines two key environmental arguments. He points out that they can be found in Rawls' *Political Liberalism*. They are:

(a) The set of political values which are not drawn from comprehensive doctrines but can be accepted by all reasonable citizens because they are neutral among reasonable doctrines;
(b) The comprehensive arguments or those arguments that are drawn from particular comprehensive doctrines.

The first kind of argument is political and is not drawn from comprehensive world-views. The second kind of argument for environmental awareness emanates from comprehensive moral doctrines which may win out in democratic deliberation. The second type of argument is more controversial however. The environmental philosopher, Bell, outlines Rawls' general approach. Bell summarises Rawls' argument:

> If something is a 'good, politically speaking,' it makes a positive contribution to the maintenance of a co-operative society of free and equal citizens, each with the capacity to form, revise, and pursue their own doctrines, and the ability to live by principles of justice appropriate for such a society.[19]

However, history has shown that principles of justice vary among communities as much as cultural and social values. Where societies potentially come into conflict is where the values of one community collide with those of another. Thus, the first key Rawlsian environmental argument emanates from his work on primary goods. When Rawls first derived these principles the environment was not a primary concern of the marketplace, and his two principles of justice therefore focussed on *human* relationships in a civil society. It is ironical that if the environment is a primary good then it should have been recognised in Rawls' first work.[20] Accordingly, in 1993 Rawls modified his stance to demonstrate that liberalism could in fact respond by accommodating growing environmental concerns. In *Political Liberalism* he said:

> [T]here are numerous political values here to invoke: to further the good of ourselves and future generations by preserving the natural order and its life-sustaining properties; to foster species of animals and plants for the sake of biological and medical knowledge with its potential applications to human health; to protect the beauties of nature for purposes of public recreation and the pleasures of a deeper understanding of the world.[21]

Within this section of political liberalism are three arguments. The first two are about polluting the world until it cannot be habited and the need to maintain a co-operative society. The third argument involves protecting the beauties of nature for purposes of public recreation – again this line of reasoning focuses on our primary goods which are the goods that we may want whatever else we desire.

Obviously these three examples illustrate the limits of political reasoning. This is because each example is essentially stemming from anthropocentric quarters. Thus, the second Rawlsian strategy is to explore the role of comprehensive moral doctrines. More particularly, Rawls's revisions to political liberalism and concessions to comprehensive doctrines continue to be shaped by the current legislative and social structures. Rawls states:

[A] bill may come before the legislature that allots public funds to preserve the beauty of nature in certain places (national parks and wilderness areas). While some arguments in favour may rest on political values, say the benefits of these areas as places of general recreation, political liberalism with its idea of public reason does not rule out as a reason the beauty of nature as such or the good of wildlife achieved by protecting its habitat. With the constitutional essentials all firmly in place, these matters may appropriately be put to a vote.[22]

It is worth remembering that in liberalism there are competing notions and arguments involving the 'the right' and 'the good'. The good is concerned with an ordered scheme of final ends and by extension the type of world in which we would want to live. Principles of right, by way of contrast, involve the derivation of principles of justice. Rawls' liberalism develops the principles in a framework that determines the ordered scheme of final ends.

Therefore Rawls moved clearly away from ontology to advocacy but left open ontological thinking to supplement his work. The relevance of Rawls' political liberalism is significant because he releases liberalism from its moorings in negative liberty and a minimal state.[23] Proponents of liberalism argue that it is capable of 'refinement' and can 'evolve' to emerging and significant issues even though these revisions have created a pastiche of ideas. In this regard, even Taylor seems to agree, saying that '[e]valuation is such that there is always room for re-evaluation'.[24]

However, where the liberal and interpretivist schools of thought differ is that the latter claims to delve into the conditions of our being-in-the-world thereby putting us in touch with nature's meaning. Furthermore, Taylor continues that procedural liberalism does not explore how environmental agreements are to be reached. The liberal response has been to note that liberal societies are not necessarily in conflict with green values and there is no necessary contradiction between liberal neutrality and its democratic structure. The environmental issue however is whether this is enough to come to terms with the contingencies posed by the natural environment.

7.4   Taylor's political approach to environmental politics

Central to Taylor's argument is that developments in modern liberalism offer a narrow approach to politics and do not examine our place in the natural environment. According to Taylor, procedural liberalism submerges the analysis of good structures which are relevant to the politics of the environment. A liberal focus on rights and individual personal freedom means that people look inward rather than to external things, such as the *polis* or the republic, and even less to the environment. The problem for democratic liberalism is that not all environmental values and voices can be heard equally. A decision in favour of development obviously overrules the environmental objections of those who oppose it. This issue reflects the pragmatic mantra that it is not possible to focus on all people's values at the same time. This maxim, while lacking Kantian categorical force, could be used to argue that the dominant voices submerge other ways of being-in-the-world. The key problem is to determine when should the state provide special treatment for environmental values which might conflict with the need for all citizens' voices to be expressed and not just periodically when elections are held.

I use Taylor's work to question whether liberalism can solve environmental issues by extending rights to accommodate what are essentially issues of significance. These recent developments in extending liberalism presuppose that the current system is adaptable and amenable to change. Taylor argues that the liberal solution only offers a superficial understanding of reconciliation, even if liberalism is defined more broadly. It has not begun the process of reconciling these differences on terms that both sides of the dispute can understand. Taylor's central critique of liberalism can be found in his examination of cultural politics. I have argued that our cultural and economic dispositions toward the natural environment are critical issues. Long ago, Taylor commented on Kymlicka's *Liberalism, Community and Culture* which seemed to reduce culture – which is a vital constituent of identity – simply to a resource:

What has happened is that a proponent of liberalism has found his own reasons to support the struggle for cultural integrity of some peoples, who for their part are engaged in this struggle for their reasons. The two are not the same. The liberal accords a culture value as the only common resource of its kind available for the group in question. It is the only available medium for its members to become aware of the options. If these same individuals could dispose of another such medium, then the case for defending the culture would evaporate. For the people concerned, their way of life is a good worth preserving; indeed, it is something invaluable and irreplaceable, not just in the absence of an alternative, but even if alternatives are available.[25]

Taylor insists that procedural liberals only *manage* the demands of cultural and environmental issues through procedural recognition. His central argument is that this liberal strategy is scarcely enough to come to terms with the type of understanding derived from an exploration of the ontology of identity. Accordingly, Taylor argues that Kymlicka's liberalism grants culture procedural recognition but not a full appreciation concerning its impact on peoples' being and the good society. For example, in the context of cultural politics there is no acknowledgement that culture is a significant good, which needs recognition on its own terms. Taylor questions whether this liberal political approach can integrate significant values – hyper-goods – into the basic institutions of liberal democracies and develop our appreciation of environmental politics.

Considerations of good political structures which might be environmentally embedded have been displaced by considerations of principles of right behaviour. Taylor has addressed the issue of equal voices:

Another obvious difficulty for a liberal defense of community aspirations is the element of coercion and discrimination it may unavoidably entail. Suppose a minority of members want to assimilate. But suppose, at the same time, that the laws needed for the survival of the community require everyone's support. So all members of the community are taxed, say, to support some institutions; or in the name of survival, which a minority are happy to forgo, their choice of language is restricted. Once we get caught in the rigid requirement of equal concern for everyone's view of the good, are we not forced here to violate our principle [of equality]?[26]

This issue applies to environmental politics which is addressed in the interpretivist emphasis on social frameworks that inform citizens about these dilemmas. This involves accommodating two critical features of deliberation. First, citizens have opportunities to participate in the formation of the good, although what seems to be a primary or common good may change over time with the will of the people. Second, citizens may find themselves in agreement with a government or state on some points, and not on others. For Taylor, if political theory is to assist the collective dimensions of community structures, there is a role for the state in creating conditions for dialogic interaction to arrive at recognition of significant goods such as environmental preservation.

Yet, simply bringing an issue into the societal structure according to liberal principles offers only procedural respect. One of Taylor's themes is that the granting of cultural or environmental rights does not come to terms with how these values shape peoples' identity. Therefore, the respect granted is limited by the contours and forces of the pre-existing system. These limitations submerge not only creative solutions but the role for perceiving and feeling the external values that are contained in the natural environment. He argues that this dilemma requires the exploration of how external values are a commonality that must be incorporated into deliberative processes. Alternatively, such external values may be accommodated to the extent that the system allows for those additional values, even if they are not commonly appreciated. It has been argued that the natural environment contains intrinsic value but that this awareness has been submerged in the endeavour to control the natural environment using economic and procedural methodologies. Taylor's work is relevant in providing means to determine whether liberal environmentalism can recognise the natural environment as a strongly valued good. In *Sources of the Self,* he develops the idea of a hyper-good which are those goods that determine the outcome of other goods and may not have a market value. A hyper-good is explained by Taylor in the following way:

[T]he picture of moral life in which a hypergood figures is one where we are capable of growth from a 'normal' or 'original', or 'primitive', or 'average' condition, in which we acknowledge and orient ourselves by a certain range of goods, to a recognition of a good which has incomparably greater dignity than these. Our acceptance and love of this good makes us re-evaluate the goods of the original range. We judge them differently and perhaps experience them quite differently, to the point of possible indifference and, in some cases, rejection.[27]

Taylor's idea is that hyper-goods are supreme among strongly valued goods in an individual's moral framework. Abbey explains these goods are capable of commanding even more respect and admiration than other strongly valued goods and can become hegemonic in a person's life.

For Taylor, however, this analysis of hyper-goods is not enough to explore the problems of modernity. He takes the argument further to explore how our secular age has ignored other critical and interpretivist ways to think about the natural environment. He continues this critique in the following way because the rise of modernity is not just a story of loss or of subtraction.[28] 'Because we can certainly go on experiencing fullness as a gift from God, even in a disenchanted world, a secular society, and a post-cosmic universe. In order to be able not to, we need an alternative.'[29]

On Taylor's view it is important to understand how we have moved between enchanted visions of the universe and a prevailing disenchanted world-view. This argument leads to the main point of his recent work: 'that a secularist régime must manage the religious and metaphysical diversity of views (including non- and anti-religious views) fairly and democratically'.[30] This task might include setting certain limits to environmental, religious and spiritual actions in the public sphere.

As Taylor points out democratic regimes also involve placing limits on non-religious or anti-religious philosophies (which also promote environmental values). From earlier chapters it is important to remember that religion is not the primary focus of secularism because Taylor's main aim is to re-enchant the system so that people may recognise the importance of the natural environment. In this way, our horizons of value are broadened and the role of public institutions developed to accommodate an appreciation of these dilemmas. For this to occur, not only do environmental concerns need to be considered within a society's collective thinking, but environmental values must be recognised in the deliberative process.

However, when a social system has external elements imposed upon it, or presented to it, to which it has no natural affinity and/or no pre-existing mechanism by which those elements may be introduced and embraced, conflict invariably is the end result. The issue for modern democratic theory is how political systems can accommodate values of significance which include environmental factors. Interpretivists offer a different way to think about our being-in-the-world and are optimistic in their attempts to re-enchant our thinking. On Taylor's view, we do not merely need more liberal procedure and procedural respect. Rather, we need an ontological and advocacy investigation into political pathways that put us in touch with the natural environment. His interpretivist work offers answers to these questions through an exploration into the connections between humanity and the natural environment. Taylor's critique of neutral liberalism and its attempt to persuade fellow citizens about the value of their green ideals might not be enough. His interpretivist thought takes us further than liberal extensionism to offer answers to these questions through an exploration into the connections between humanity and the natural environment. A further problem for procedural forms of liberalism is the reliance on instrumental reasoning as utilised by economic technocrats to arrive at maximal outputs at minimal costs.

7.5   John Dunne and Taylorian resources: some democratic responses to the environmental dilemma

This section offers some speculative thoughts concerning how we might overcome some of the environmental problems that plague modern liberal democracies. The political philosopher John Dunn explains the key problem:

> As of now, we have very little grip on the central question of just what we can trust modern representative capitalist democracy to do on our behalf. This is more alarming if we regard it, as on present record we surely must, as congenitally indiscreet and often wantonly harmful to the environment.[31]

Invariably, representative systems and methodological individualism take us back to the free market and approaches that assume sustainable futures are possible. I argue that Taylor's interpretivism offers a critical environmental ethic in a broader space of reasoning with democratic overlays.[32] Taylor's work implies that we need to totally rethink our basic set of assumptions and directions for our communities. He has argued that a collectivist approach focuses on common bonds that can provide workable environmental solutions. The environmental issue is not simply about justice, but creating the preconditions to recognise that the natural environment is a necessary condition of our being.

I have argued that Taylor is highly relevant in regards to ecological politics. Obviously, in this respect the environmental issue poses a key problem for modern communities, because at times environmental concerns may be in direct conflict with economic imperatives and realities. Perhaps the issue is resolved ultimately only when all members of the community share this common endeavour, and the idea of shared social understanding becomes more pressing for democratic theory. In early political theory it was assumed that the mechanisms of power such as an absolute monarchy or a benevolent dictatorship were the means to create and nurture a general will. From Taylor's work on identity and secularism, the key issue involves the development of deliberative structures throughout society to critically analyse, inform and nurture the significant values that shape our being-in-the-world.

The analysis of single-focussed and procedurally focussed governments is far removed from the liberal-democratic model that is investigated in this chapter. What environmental issues demand is a developed, deliberative and democratic model that is able to accommodate and consider how significant values are to be determined, which might create commonalities involving nature's value. At the very least, Taylor's exploration into the possibilities of our being-in-the-world offers steps toward the creation of new directions for modern communities to consider in an ecological context. Again, it is in John Dunne's work that a useful summary of the overall inherent contradictions of democracy and liberalism can be found. He explains:

Democracy, in any understanding, has always been a singularly weak conceptual candidate for specifying deliberative rationality. It may, as Aristotle conceded, have the advantage of preventing the exclusion of potentially relevant considerations, but it conscientiously refrains from imposing any other constraints on deliberative outcomes. Moreover, it does nothing to clarify which reasons are better, and should therefore carry greater force, and which are more alluring and popular and hence carry greater immediate power.[33]

From democratic to liberal theory there does not exist a fully worked out solution to the dilemma concerning how significant environmental values, such as the supposition that nature is a necessary condition of our being-in-the-world, are integrated into the polity. Authenticity is a key concept in Taylor's work, defined in one sense as uniqueness, but more generally as autonomy within context, where the 'context' is not viewed in instrumental terms. Taylor's interpretivism is about reminding us that authenticity may not be weighed nor traded because it is about being, rather than choosing. Taylor here echoes Heidegger's view that life is a current of being and authenticity a political goal. Authenticity is pursued through a process of practical reasoning in a political context where there is equal facilitation of, and access to, debate in democratic public spheres.[34] Taylor argues that it is through language that practical reasoning exposes new and different ways of living and relating with the world. In particular, it is important to consider the relationships between environmental considerations and social factors such as class, regional autonomy, nationalism, and identity. Through discursive public spheres, Taylor's first step involves appreciating the views of others as they shape the collective good: the voice of nature reflects the myriad interconnections between people and the natural environment. As Taylor put it:

> It is just conceivable that some will arise which will themselves prove superior, more likely there will be some where reason cannot arbitrate, and almost certain that we pay a price for our universalism in the loss of some goods which were bound up with earlier, more encapsulated forms of life. But none of this gives us cause a priori to take refuge in an agnostic relativism.[35]

One task for the politicisation of environmental values, then, is to provide opportunities for citizens to participate actively in the decision-making processes of communities. Taylor's aim is to respond to critics who have argued that his work on secularism is derived from a Western point of view and he does this by broadening liberal conceptions of legitimacy. For example, critics often point out that developing economies do not need simply principles of justice but access to resources and other factors which would lead to the development of their economy. Central to Taylor's evaluative framework is the argument that modern liberal society appears as a collection of citizens without place of origin that have banded together for their common protection. Yet Taylor, not as pessimistic as Dunne, operates with the spirit of humanity which is implicit in *A Secular Age,* where the public sphere of our modern age is contrasted with the public sphere of the ancient republic, or polis. His aim is to create new discursive, political and social pathways that open us to imaginative explanations concerning relationships with the natural environment.

In *A Secular Age,* Taylor emphasises the power within communities that can ultimately remove governing bodies that stifle the human condition. His purpose is to recall how these ancient models included discussion outside the decision-making bodies of governance. Through improved interpretations and improved discursive structures it is possible to subvert the current order of things as they apply to environmental politics. The difference between the ancient polis and the public spheres of today is that in the ancient polis people could hold governments to account. Taylor explains:

> Now there is a subtle but important difference. Let's compare the modern society with a public sphere with an ancient republic or

polis. In this latter, we can imagine that debate on public affairs may be carried on in a host of settings: among friends, at a symposium, between those who meet in the agora, and then of course in the ekklesia where the thing is finally decided. The debate swirls around and ultimately reaches its conclusion in the competent decision-making body. Now the difference is that the discussions outside this body prepare for the action ultimately taken by the same people within it. The 'unofficial' discussions are not separated off, given a status of their own, and seen to constitute a kind of meta-topical space. But that is what happens with the modern public sphere. It is a space of discussion which is self-consciously seen as being outside power. It is supposed to be listened to by power, but it is not itself an exercise of power.[36]

In the modern public sphere these debates are being seen as outside of power. With the rise of the modern public sphere comes the idea that there is a need for political power to be checked because of the separation from the people. It is in this sense extra-political. *A Secular Age* contains a perspective on language which provides better interpretations to adjudicate between different narratives. With typical optimism, improved interpretations act to moderate the excessive scope bureaucratic reason has in the modern world.

These interpretivist ideas are not about idealising the public sphere but engaging with some of the dilemmas of capitalism that have led to environmental and social inequalities. Taylor's argument for change through the public sphere is based on the view that the economic and material causes of our current dilemmas can be moderated by an explicit political approach that investigates the *ideas* behind them. After all, it is human communities that have created the economic systems that divorce people from each other and their natural environments. According to Taylor, the public sphere is the arena to reveal the moral worth of different visions of the good life, which includes our interactions with the natural environment.

7.6 Conclusion

This chapter examined Taylor's criticism of liberalism to explore environmental values as a source of the self. The value of liberalism is, of course, its emphasis on the rule of law and its sincere commitment to the protection of fundamental individual rights, reflecting certain common features among all people. Its shortcomings have been its proceduralism, instrumentalism and the priority it gives to negative freedom without exploring what is significant for identity. Taylor's interpretivism may supplement liberalism in these respects and particularly by casting freedom in a more positive way. That is, as self-determination which involves a role for the state by recognising differences and environmental values.

It is argued that Rawls endorses negative freedom and is somewhat suspicious of the state. He has been criticised, moreover, for offering us no theory of the state. Nevertheless, if limits to growth become real, Rawls' principles of justice demand reform to accommodate environmental values. The state is vital in providing a democratic forum within which to foster ecologically aware politics which will not emerge spontaneously but must be forged by democratic institutions. Taylor's work emphasises a role for the state that involves the public sphere, and involves more than state directives for sustainable futures. It involves shaping our civil societies in a manner that counters both corporate power and excessive state power in policy.

In this regard, Taylor's work was used to develop Rawlsian constructivism to provide new ways to reform existing basic institutions. He argues Rawls' political liberalism orders difference but lacks political engagement with others who hold different legitimacy systems to those we have in our secular age. Taylor extends the ideas of Hegel, who was interested in exploring 'the spirit of humanity', while struggling toward an awareness of humanity's 'being-in-the-world'.[37] Therefore, Taylor's work on environmentalism and secularism is to recognise that there is more to people's identity than simply following a liberal procedure and rule. The ecological implication of Taylor's interpretivism is that environmental values must be implemented through advocacy of a politicised civil society. That is, where the state and its institutions influence but do not determine how people are to live their lives. These visions for an environmental world are guided by a normative commitment which respects the value in the natural environment.

Taylor's view is that procedure simply orders differences without fully coming to grips with the underlying dilemmas that they represent. He is particularly critical of representative systems that limit the means through which we can imagine and develop a new conception of ecological politics. More fundamentally, a procedural system has the potential to create a bureaucratic structure from which we cannot escape. As an example of the type of democratic reforms, Taylor suggests a return to an earlier way of moderating power in a public forum, the ekklesia, which would be called for in the Greek city states when their governments had become too corrupt and oppressive. This was an assembly outside the civil authority of the city and if enough people came out and refused to accept the existing centralised civil authority that government would collapse. Non-participation has been a successful and peaceful means to free mankind from oppressive civil authority throughout history.

8
Interpretation, Language and Environmental Values: The Habermas and Taylor Debate

This chapter examines the environmental relationships in civil society by using the debate between Taylor and Jürgen Habermas. The aim is to demonstrate how environmental commonalities and differences can be reconciled in new political spaces. Taylor's view of language has been criticised not only by Habermas but also by a number of commentators. They are the focus of the final section of this chapter.

8.1 Introduction

This chapter examines Taylor's political work on language, intrinsic value and meaning in the natural environment.[1] In his classic book, *Hegel*, he analysed the conflict between unlimited economic growth and the preservation of the natural environment. He often commented in his early work that environmental dilemmas were likely to emerge because of industrialisation, which was a focus of his essay 'The Politics of the Steady-State'.[2] In his most recent work, he suggests political approaches that emphasise the role of language to reconnect us with the natural environment which reflects the meaning and value in it. In 'Heidegger, Language and Ecology'[3] Taylor locates important insights for culture and the natural environment in the various works of Hamann, Gadamer, Heidegger, Herder and Humboldt. He uses ideas from these theorists to explore connections between our language and the natural environment, which led to his argument that nature is a source of the self.[4]

This chapter is structured in five key sections following this introduction. Section 8.2 introduces Habermas's criticisms of Taylor concerning the importance of ecological values. Section 8.3 outlines Taylor's criticism of Habermas's approach to language and meaning. Section 8.4 explores recent debates between Habermas and Taylor concerning public reasons. Section 8.5 responds to some recent criticism. Section 8.6 concludes the chapter.

8.2 Habermas's criticisms of Taylor's ecological approach

Critics such as Jürgen Habermas claim that Taylor's arguments are based on spurious metaphysical claims with unforeseen political and religious consequences. These critics claim that Taylor's emphasis on connections between meaning and intrinsic value could lead to animal and natural rights taking precedence over human rights. Here Jürgen Habermas enters the debate. He engages Taylor's quest to open us to the otherness of the cosmos and the natural environment. Habermas argues that this focus leads Taylor to support external values. Habermas argues that cultural 'survivance' is at the expense of individual rights. Habermas explains:

> The ecological perspective on species preservation cannot be transferred to cultures. Cultural heritages and the forms of life articulated in them normally reproduce themselves by convincing those whose personal structures they shape, that is, by motivating them to appropriate productively and continue the traditions. The constitutional state can make this hermeneutic achievement of the cultural reproduction of the life-worlds possible, but it cannot guarantee it. That is, to guarantee survival would necessarily rob the members of the very freedom to say yes or no that is necessary if they are to appropriate and preserve their cultural heritage. When a culture has become reflexive, the only traditions and forms of life that can sustain themselves are those that bind their members while at the same time subjecting themselves to critical examination and leaving later generations the option of learning from other traditions or converting and setting out for other shores.[5]

For Habermas, it is the search for universal principles in the chaotic and ultimately indeterminate universe that must be the focus of politics. Here, it is worth remembering John Dryzek's succinct summary of the limitations of Habermasian discourse in which he argues that the limitations arise from 'what bourgeois society currently allows and encourages in the way of individual development'.[6] These issues can be addressed when we think about language as a common and shared endeavour where no-one is the master in the ebbs and flows of conversation. No fixed solutions are assumed because the outcomes of dialogue are always uncertain where a deeper appreciation of the other is assumed to follow through continual interaction in discourse.

8.3   Taylor's criticism of Habermas's approach to language and politics

Central to Habermas's argument is the claim that Taylor's strong hermeneutic framework is built on a foundation pioneered by Herder, Humboldt, and Heidegger.[7] Accordingly, it is Habermas's argument which perpetuates an unjust political structure that condemns people to the daily grind of coping; that is, life is defined according to the parameters of a procedural rule without a vision about what it is good to be.[8] Taylor engages with Habermas to provide deeper argument and engagement with the natural environment. In this debate with Habermas, he argues that in the work of Kant we can trace back a major line of thinking that divided our minds from the natural environment. This separation creates a disenchanted understanding of the natural environment: Habermas's procedural model remains trapped in the mind and the world dualism, which was the focus of Chapters 3 and 4.

Central to Taylor's framework is his ethic of authenticity which takes issue with the argument that procedural rules reflect the spirit of liberalism but fails to adequately explore the potential of dialogic and expressive models.[9] The modern liberal infatuation with procedure does not explore that common ground where differences are respected in a spirit of openness and democratic healing.[10] For Taylor, the common good reflects that political space where agreement is the fruit of democratic interchange and discourse; namely, it is that space where people's strong values are articulated, expanded and sometimes challenged in the quest for justice.[11] Justice might mean that one group gives more ground on a certain issue, owing to their erroneous stance on a particular matter.[12] More particularly, he dismantles the system-building intent of procedural liberal theory and advocates a republican and discursive democratic model whereby an element of the universal in a particular linguistic phenomenon is retained. Taylor's work on language and discourse creates workable political relationships between particular cultures by locating what is common between them; this argument is based on the political projection that differences can flourish in a tolerant community.[13] It is therefore a category error to argue that Taylor ignores the dimensions of all that are different and diverse in society by focussing on what is common. Rather, his focus is on the problems of identity which haunt modern politics, and he exposes the illusions and pretensions that a fixed yardstick offers. Taylor's solution is based on his conception of practical reason which squarely faces concerns of commonality, difference, foundationalism and proceduralism.[14]

Briefly, then, the differences between Habermas and Taylor involve the intersubjective and objective role that is performed by language; that is, it involves understanding how conversation, discourse and language cultivate the possibilities for a good life.[15] Taylor's political strategy is cultivated within a civilising society that is committed to moral progress and development.[16] Intriguingly, their very different proposals are based on the work of Herder and Humboldt even though they project different solutions and proposals. As mentioned earlier in the introduction to this chapter, Habermas maintains that language offers a unique opportunity to construct a decision model where differences can be solved once and for all by setting and applying a rule in the discourse arena.[17] Here, the purpose of Habermas's model is to *assess* communication that is incompatible with the good or better life. Habermas sets a procedural test that substantive claims must pass in order to be normatively valid. This test consists in universal rules of discourse – reciprocal accountability, inclusiveness, freedom to question claims and to presuppose counter-claims, and non-coercion.[18] These reflect the legitimate procedural constraints that we are entitled to make, that indicate the basis for rational agreement regarding the 'justness' of a given norm, and of assertoric statements. Taylor, however, responds with the argument that language discloses new worlds through an ongoing conversation of humanity and opens up new attitudes towards the other.[19]

For his part, Taylor believes that Habermas ignores the nexus between the structure and practice of language and thereby seriously undertheorises the background contexts and nuances which shape people's very being-in-the-world: where the nuances of culture and the life-world impact on the formation of a language. Habermas responds and maintains that:

Of course Humboldt is not an empiricist trying to pull the rug out from under the feet of the process of reaching understanding and hold the identity of linguistic meanings to emanate from the randomly iterated intentions – constantly superseding each other – of single, isolated speakers. For him, the intersubjectivity of a common perspective does not dissolve, for example, into a series of isolated

I-perspectives which are merely reflected in one another; rather, it arises at the same time; and from the same source as intersubjective validity of semantically identical linguistic expressions and is of equal origin (*gleich-ursprunglich*).[20]

Habermas argues that Taylor's Hegelian focus is on the 'we' perspective implicit in language, which leads to the socialisation of individuals within practices that abrogate personal rights.[21] As stated above, these set a procedural test which substantive claims must pass in order to be normatively valid.[22] They consist in the universal rules of discourse which reflect the problem-solving power of language, and these rules are at the disposal of each individual agent to affirm or reject the norms and expectations.[23]

Habermas argues that Taylor has lost sight of the problem-solving capacity provided by language and thereby overplays its world disclosing possibilities. For Habermas, Taylor's broad and interpretive conception of language submerges an appreciation about what language is and thereby also submerges the substantive reasons that shape the good life. This may be taken as a generalisation which is itself a reflection of Habermas's argument that Taylor is too quick with his introduction of philosophical ethics into the equation, thereby ignoring the previously mentioned problem-solving possibilities opened up by language.[24] The procedural test, according to Habermas, is constructed from within the intersubjective capacities of language and is not exhausted by the individual and communal dimensions of language.

Large differences between Habermas and Taylor clearly remain and involve the role that language plays in human affairs and in constructing communities. In Habermas's view the problem-solving possibilities implicit in discourse are lost when practical reasoning is given full scope, and as a result Taylor's framework ends up justifying a singularly focussed common good. However, Taylor's notion of commonality involves a conversation of humanity where the politics of the 'good' develops out of workable political arrangements that require careful political manoeuvring. Taylor continues by arguing that Habermas's ethic tends to create a subtle and unacknowledged suppression of difference; that is, by developing more procedural rules we do not provide any convincing substantive reasons why people should give ground and live their lives according to that rule. A simple focus on procedure, at the expense of interpretation, is too abstract and leaves the associated identity issues untouched. Here, Taylor is at his most rhetorical in his argument that it is our being, culture, and language which shape the creation of a rights culture and this is something that can never be reduced to procedure. Taylor claims that Habermas's theory of language is too scientific in assuming that words are precise devices at the disposal of interlocutors and Habermas therefore misses the art of disclosure. Habermas assumes:

> that all participants involved use the same expressions in a semantically identical manner, for, without this idealizing precondition, they would not be able to enter into dialogue in the first place.[25]

Habermas argues that words act as technical and problem-solving devices. It is entirely problematic, however, whether the role of language may be idealised in Habermas's sense. The problem that Taylor identifies in this approach is that it leaves the source of conflict in the 'life-world' untouched. Therefore, Taylor claims that conversation and dialogue do not simply reflect the structures and rules that give words meaning, nuances and often ambiguities. For Taylor, this strategy loses sight of the possibility that within the ebb and flow of conversation an element of the universal exists within particular speech. Words are not precise devices at the hands of an interlocutor in an Enlightenment sense. This is not necessarily a problem in Taylor's argument because the discourse that uses these nuanced expressions leads to new ideas and concepts.[26] Moreover, these concepts must be considered in the light of the place and context to which they belong.[27] For Taylor, therefore, language and expression cannot be tamed by Habermasian inspired Kantian procedure.[28] Consequently, it is problematic whether a set of rules cultivates character or, instead, frustrates citizens in that they must always follow a rule to which they are not accustomed.[29] According to Taylor, Habermas's theory generates anomie, frustration and ultimately perhaps a recalcitrant polity.[30] For Habermas, on the other hand, a clear and precise solution must be found in language.[31] Habermas's reliance on Kantian procedure[32] blocks the power of language, and its individualism prevents developing that common space where people respect others and the differences which make up society.[33] Taylor argues:

> So Kant's thesis [is] that moral reasoning imposes on us the requirement of being able to universalise the maxims of our will, and Jurgen Habermas' discourse ethic is seen as binding on us in virtue of our being interlocutors who seek to convince each other. Now this too doesn't seem to me to work. It may seem to leave us without any way of backing our feeling, that we have come to an ethically superior

position in relation to our ancestors of 400 years ago, with reasons. But I don't believe this is so. There is a way of proceeding, by what I have tried to call 'supersession arguments', where we show that there is a rational path from A to B, but not in the reverse direction. But we have to see this path not only as a line of argument, and not only as an actual transition, but as both together.[34]

In this passage Taylor's enthusiasm for reforming modernity reflects his explanation that a full and fair hearing is a necessary precondition for a fair settlement. For Taylor, Habermas's solution cannot reconcile differences between people where the procedural test necessarily excludes people's strong evaluations.[35] The problems of assimilation, culture and reconciliation become even more critical when Habermas's procedural language methods turn back to naturalistic and scientific interpretations of the life-world. Habermas's tendency to prioritise scientific and precise solutions through the craft of language can never be created with scientific precision; after all, it is politics and discourse which reflect the twists and turns of a full life. Words and conversation have different meanings in various circumstances, which reflect the ebbs and flows of the world. To assume that language is a precise tool in the hands of each person underestimates the reciprocal dimensions of language; that is, freedom also involves 'recognition of another agent of it as binding on both of them'.[36]

For Habermas, the only role for moral theory is to single out a procedure to judge the moral worth of particular claims.[37] In this manner he shifts the Kantian injunction to universalise our moral maxims to the individual participants in the discourse process itself. This reflects Humboldt's assumption that '[o]nly in the individual does language attain its final distinctness'[38] where the capacity to express thoughts through language is at the possession of each individual actor. The next move Habermas makes is to argue that this possession lies in each person's mind such that he has the capacity to arrive at fair resolutions to significant matters on his own. Taylor's response is that language is not simply a resource for each person, but reflects a seamless web of interconnections that connect people with shared purposes. Language is not something at the disposal of each person, but reflects a shared conversation of humanity through dialogue. For his part, Habermas's criticism is that Taylor's expressivism conflates different values within this common space and ignores people's rights to individuality. Habermas maintains the politics of the good leads to political problems arising from the relationship between 'strong reconciliation' and claims to 'self-interest'. The search for common purposes and a stronger conception of reconcilia-tion, according to Habermas, leads not only to value scepticism, but also to the inherent politics of the good. Habermas continues:

> What moral *theory* can do and should be trusted to do is to clarify the universal core of our moral intuitions and thereby to refute value scepticism. What it cannot do is make any kind of substantive contri-bution. By singling out a procedure of decision making, it seeks to make room for those involved, who must find answers on their own to the moral-practical issues that come at them, or are imposed on them, with objective historical force.[39]

For Habermas, moral theory is used as a base for the attainment of a social ethic which is *singularly* practical and problem-solving in its implications. This sociological generalisation is invested with more philosophical weight than is justified.[40] That is, it seems to ignore the deeper sense of reconciliation involved in a substantive conception of the common good which brings people together in a spirit of recon-ciliation that entails embracing and engaging others who are different from us. At this point, Taylor develops Hans Georg Gadamer's work on horizons of meaning that shape our lives and involves repairing and healing our divisions while correcting the errors in interpreting the other.[41] Nevertheless, Habermas's theory of language is a procedural epistemology which, with its associated theory of justice, offers only a surface-level understanding of the motivations that shape what it is to be a person.[42]

For his part, Habermas responds by arguing that Taylor's ethic and his search for reconciliation leads to relativism and cultural disharmony. Taylor disputes these concerns with relativism and takes the debate back to the language work of Herder and Humboldt; which involves deter-mining whether the role of language theory is predicated on a theory of mind where words precisely express each person's thoughts.[43] Rather, language is a common possession shared between people and the use of language creates a richer political process that cannot be reduced to procedural rules. For Taylor, politics involves teasing out the differences which are reflected in discourse and which for him involves error and misinterpretation. Taylor summarises this point when he states that:

> [I]t seems to me indisputable that a similarly substantialist theory is also necessary in the case of such a reinterpretation – and a purely procedural ethics thus radically misses its goal.[44]

In a more recent response, Habermas argues that the discourse principle applies only when 'real' individuals affirm the ethical discourse principle.[45] Arguably, this reveals a paradox in the structures of Habermas's reasoning. This paradox is that universal moral maxims cannot be implemented unless they are affirmed and developed by participants in real world politics. Norms and assertions therefore are no longer universal in scope.[46] Taylor's dialogic approach to politics assumes that language cannot be mastered by any one interlocutor standing apart from others; nor can Habermas rely on scientific methods in dealing with human affairs.[47] Taylor, therefore, expresses grave reservations with Habermas's designative, procedural and scientific conception of language which leaves differences intact.

Therefore, Taylor's recommended approach to politics is not a blueprint, but a tentative framework with an associative and interactive model which presupposes engagement in discourse.[48] According to Taylor, it is Habermas's model which is too abstract in that it leaves the identity of participants and the sources of conflict originating in the life-world untouched.[49] For Taylor, the political strategy to solve the commonality-difference conundrum involves recognising local and context specific factors as they create and work to achieve a good society based on that which is common between people.[50] Of course, ideas drawn from an expressive interpretation of the role of language are not designed as a critique of democratic theory; and, it is of course right that public rules lean heavily towards the protection of individual rights. As stated, it is through language that shared experiences are created where no-one is the master in the ebbs and flows of conversation. No fixed solutions are assumed because the outcomes of dialogue are always uncertain where a deeper appreciation of the other is assumed to follow through the continual interaction in discourse.

8.4   Habermas and Taylor: recent debates over public reasons

As mentioned earlier Taylor's work on Herder, Humboldt and Heidegger is the focus of Habermas' critique.[51] According to Habermas, the post-liberalism (interpretivism) of Taylor moves too fast in that it relies on philosophical ethics at the expense of language as a problem-solving structure; that is, Taylor overemphasises the world disclosure function of language and fails to adjudicate between different political claims.[52] For critical researchers, Habermas invites us to consider the motivations that led us towards the neutral decision model – the ideal speech situation.

Habermas continues that there is no reason to oppose one sort of reason (secular against religious reasons or Hegelian or Kantian) simply because many people find religious reasons are coming out of a world-view which is inherently irrational.[53] However one of the aims of Taylor's work is to open us to different world-views such that we understand how our common human reason works in environmental awareness and religious traditions. Of course Habermas is correct that religious, secular or any other cultural enterprise should not be given precedence in the public sphere. Habermas continues:

> However, if it comes to lumping together Kantianism and utilitarianism, Hegelianism and so on with religious doctrines, then I would say there are differences in kind between reasons. One way to put it is that 'secular' reasons can be expressed in 'public', or generally shared, language. This is the conventional sense that Chuck [Taylor] is trying to circumvent by introducing the term *official* language. Anyhow, secular reasons, in this sense belong to a context of assumptions – in this case to a philosophical approach, which is distinguished from any kind of religious tradition by the fact that it doesn't require membership in a community of believers. By using any kind of religious reasons, you are implicitly appealing to membership in a corresponding religious community. Only if one is a member and can speak in the first person from within a particular religious tradition does one share a specific kind of experience on which religious convictions and reasons depend.[54]

Arguably, Habermas is avoiding Taylor's long-standing criticism that he is ranking differences and not reconciling them. Habermas responds that the key problem with the Taylorian interpretivist approach is that it seems to emphasise participation in cultic and religious practices in which reasons are tied to membership. It must be stressed that Habermas's approach to modernity emphasises the role of reason in transcending our ties to cultic, communal practices and environmental values.[55]

As the environmental philosopher Annick Hedlund-deWitt explains, the key purpose of Taylor's analysis is to open us to different world-views and their significance for global cultural and sustainability debates.[56] Habermas argues, however, that Taylor's approach to practical reason is likely to give the green light to language, environmental and group rights that would trump individual rights. The scope given to practical reason is the central difference between interpretivists and procedural

theorists such as Habermas. For Habermas, Taylor's work contains illiberal possibilities and irrational religious ideals which can be used by
cunning political actors to further their own self-interest and thereby
perpetuate political corruption.[57]

Taylor, however, has responded that his discursive language-based
politics is not an abstract blueprint, but involves normative possibilities which aim to create a life free from bureaucratic tutelage. Central
to Taylor's framework is his ethic of authenticity which takes issue with
the argument that procedural rules reflect the spirit of liberalism, but fail
to adequately explore the potential of dialogic and expressive models.[58]
The modern communicative and liberal infatuation with procedure
does not explore that common ground where differences are respected
in a spirit of openness and democratic healing.[59] The relevant idea for
environmental politics is that the common good reflects that political space where agreement is the fruit of democratic interchange and
discourse; namely, it is that space where people's strong values are articulated, expanded and sometimes challenged in the quest for justice.[60]
Justice might mean that one group gives more ground on a certain issue,
owing to their erroneous stance on a particular matter.[61] More particularly, Taylor dismantles the system-building intent of procedural liberal
theory and advocates a republican and discursive democratic model
whereby an element of the universal in a particular linguistic phenomenon is retained. Taylor's work on language and discourse aims to create
workable political relationships between particular cultures by locating
what is common between them; this argument is based on the political
projection that differences can flourish in a tolerant community.[62] It is
therefore a categorical error to argue that Taylor ignores the dimensions
of all that are different and diverse in society by focussing on what is
common.

8.5   Taylor's Interpretivism: some responses to critics

Plurality and nature

While Habermas criticises Taylor's theory of language, it is from postmodern quarters where differences are seen to be submerged in a
harmonic Taylorian view of politics. A central plank in the postmodern
critique of Taylor's post-liberalism (interpretivism) is that language does
not disclose anything in particular and all it can do is suppress differences. A suppression of difference reflects the politics of the common
good which ties people together within a confining community from
which they cannot escape.

Here work on difference is important as outlined by theorists such as Joel Anderson. From this perspective, Taylor ignores the agonism at the base of society while also neglecting the consideration of power and thereby lacking sensitivity to particular local values. Here commonality seems only illegitimately possible in such a plural and secular world. In the vast literature on postmodern ideology, it is Joel Anderson who is reflective of this way of thinking. Anderson argues:

> His [Taylor's] ontological account of non-subjective standards of value makes it difficult to envisage why the divergence of individuals' personal commitments could be viewed as a positive aspect of modern cultures in which, as Taylor consistently emphasizes, individuality and authenticity are prized. The worry remains that, without a more complete account of what it means to reason about the good *for me*, Taylor has no way of blocking the implication that, to the extent to which strong evaluators reflect critically and deliberate insightfully about how much importance certain projects, relationships, and ideals should have in their lives, there will be less and less room for reasonable divergence regarding what it is worthwhile to devote to oneself.[63]

Anderson essentially offers another sociological generalisation which is invested with more philosophical weight than it can bear. The generalisation is that it is assumed that any discussion of the common good actually betrays a strategy of the powerful to suppress the weak. It essentially assumes that an enthusiastic approach to practical reasoning has the power to create interaction that recognises people's being-in-the-world thereby opening up space for reasonable divergence regarding what it is worthwhile to be. Indeed, Taylor has stated he believes that a politics of the common good is an effective way to manage diversity. As Taylor has argued, 'if people can't come to an agreement, we will have to try something else. Notwithstanding, federalism has the advantage of providing common ground and creating more flexibility for complex identities.'[64]

Here, Taylor is essentially extending Gadamer's work of a fusion of horizons in a spirit of openness and toleration where 'through such distinctions that we can come to grasp what is distorting understanding and impeding communication'.[65] Putting Taylor's work on reform liberalism in a simple way involves the claim that the search for a common ground is about providing and creating political space where people assert their differences without fear or favour. Fundamentally, this involves the cultivation of a spirit of humanity that searches for that space where values are given full and fair expression; that is, that arena where we work to move beyond our superficial differences with a spirit cultivated by magnanimity and openness.

Unfortunately, however, this way of thinking about politics is assumed to be relativist in that the best that we can hope for is an overlapping consensus that brackets people's deeper purposes and values. These significant values are relegated to the private sphere which leads to the argument that this variant of liberalism is imperialist and intolerant of others. From Taylor's perspective the layers of meaning within liberalism are overlooked in modern political debates where the search is to create precise and technical solutions. The power of practical reasoning, cultivated through language, neglects the power that words play in creating understanding and fusing horizons. Michael Walzer's review of Taylor's broad liberalism certainly comes to grips with the discursive extensions to that doctrine. Walzer explains:

(1) The first kind of liberalism ('Liberalism 1') is committed in the strongest possible way to individual rights and, almost as a deduction from this, to a rigorously neutral state, that is, a state without cultural or religious projects or, indeed, any sort of collective goals beyond the personal freedom and the physical security, welfare and safety of its citizens.

(2) The second kind of liberalism ('Liberalism 2') allows for a state committed to the survival and flourishing of a particular nation, culture, or religion, or of a (limited) set of nations, cultures and religions – so long as the basic rights of citizens who have different commitments or no such commitments at all are protected.[66]

It is important to return to Walzer's analysis of liberalism to understand that there are commonalities within the doctrine that must be cultivated in liberalism two. That is, commonalities are needed if society is to function effectively. I have been arguing that this form of liberalism (interpretivist) opens us to differences and environmental values in a spirit of openness and security. By coming to grips with the pluralism that exists in modernity, common purposes are articulated subject, of course, to the satisfaction of basic human rights.

More particularly, both Anderson and Habermas are too quick in offering arguments that would support the supposition that Taylor would unambiguously justify environmental restrictions and language laws that would suppress difference. And even if this sociological generalisation held true it contradicts and submerges the fluid dimensions implicit in language and discourse; arguably, that would mean that we fuse horizons, confront difference, and operate on fair terms of interaction. Yet, Anderson and Habermas do not support notions of commonality which could be implemented in a federal structure that unites people despite their differences. Indeed, it is this fact that Ruth Abbey points out when she argues that a Taylorian reply involves:

[A]s immigration makes Quebec increasingly multicultural, the community of French speakers will diversify considerably compared to the image of the *'pure laine'* society prior to the Quiet Revolution of the 1960s. As a consequence, beyond the common fact of speaking French, the meanings of being Quebecois will multiply.[67]

This passage has some important implications for our environmental politics.[68] Remember that postmodern critics claim that communitarianism (interpretivism) is likely to suppress difference. However, in the above passage Taylor's interpretivism does not suggest a harmonious view of politics. The lesson from Quebec is that as immigration intensifies so will diversity. Thus, it is possible to develop the environmental implication from Taylor's work; that is, in concurring with the important argument we are able to suggest the development of commonalities broad enough to encompass a focus on difference and nature.

Difference

From another critical angle Morag Patrick, for example, has claimed that Taylor's foundational hermeneutics must be replaced by a critical hermeneutic. He argues that Taylor imposes a community structure which perpetuates the traditions and values that emanate from times gone by. Thus, Taylor's framework is idealistic and uncritical in its focus. Patrick argues:

There is a type of fallacious reasoning involved in drawing normative conclusions from ontological descriptions in this way. For even if we grant that we are engaged agents with a sense of things, this will be equally true of the procedural liberal and the classical republican. The ontological insight does not indicate that which we are to be, it is not a prescription for belief or action. As a reflexive participant I can critically examine and reinterpret facts of my culture, and it is important to stress that ontology as such cannot settle whether I am right to do so. My contention is that this critical capacity to assess, reject, *and transform* features of our common background is completely neglected by modern liberalism in that it does not confront the question of whether it is possible to arrive at a social reality that is *prior* to symbolization or representation.[69]

Taylor's hermeneutic and interpretivist framework, however, is not constructed through symbols, nor does it necessarily fail to critically evaluate other traditions. The hermeneutic dimension within Taylor's post-liberalism (interpretivism) and the attempt to fuse horizons necessarily implies that cultures change and adapt. How can horizons change if they are not subject to critical scrutiny?

Patrick's argument could be modified through consideration of Taylor's work on critical theory where he has explained that three dominant methodological traditions have commented on the dialectical method.[70] First, a scientific tradition refutes dialectical methods as mystical and subjective.[71] A second tradition offers a dialectical representation of reality, culture and nature. While a third tradition explores an ontic dialectic which investigates culture, history and politics according to notions of 'strong evaluation'.[72] Western political philosophy, according to Taylor, has taken the first direction. It rejects dialectical methods as unscientific, only mystifying social and environmental relationships.[73]

The orthodox Marxist tradition, though it more accurately follows Engels, sees the structure of reality itself as fundamentally dialectical. Schools one and two are diametrically opposed, despite the fact that they both claim to be founded on a scientific world-view.[74] The third school uses the dialectic to understand human beings, their lives and history. This way of thinking about revolutionary change implies that an ontic dialectic exists in his post-liberal (interpretivist) thought. Moreover, Taylor is at his most compelling when he argues that horizons 'change for a person who is moving'.[75] The postmodern misology that flavours Patrick's stance fails in exploring Taylor's focus on what is 'significant' thereby leaving his interpretation with a narrow means to accommodate difference. When Taylor argues that horizons change for a rational interlocutor he is actually challenging those institutions and practices that exclude, marginalise and harm others.

Therefore, this chapter contends that Taylor's work on solving the difference-commonality conundrum can be recast as a two-stage strategy to obtain a more comprehensive grasp of the assumptions, ideas and values which then frame different cultural, environmental and social frameworks. As if in response to this critique, Taylor has argued that practical arguments start off on the basis that my opponent already shares at least some of the fundamental dispositions towards good and right which guide me. Chapter 2 showed that interpretivism involves not simply hermeneutic analysis but the creation of public spheres to create face-to-face community relationships concerning better ways of living.[76] Through practical reasoning, unity and diversity are balanced in hermeneutic fashion, itself reflecting Taylor's total dissatisfaction with the alleged neutrality of liberalism and postmodern attempts at continual deconstruction.[77]

Intriguingly, Levy, Patrick and Redhead trace Taylor's error back to his interpretation of Nietzsche and the postmodern tradition. I have argued that Taylor offers a different approach to language using the very same texts from Hamann and Herder. My aim was to develop a way to think about our significant relationships which include: culture, the natural environment and religious understanding.

Taylor, nature and language: some further responses to critics

Mark Redhead has developed a unique interpretation that Taylor's critique of procedural liberalism and postmodernism is overblown.[78] This reflects his Catholic sense of the good which allegedly throws us back to a pre-modern age. Intriguingly he echoes Hittinger's observation that 'there is a theology that is crucial to, and yet left inarticulate in, Taylor's treatment of these issues'.[79] This theology, it is argued, leaves less and less room for reasonable divergence within society regarding what is worthwhile to devote to oneself. In a telling passage, Redhead argues that a 'non-ontological' account is preferable because Taylor's political theory misconceives the liberal project and actually ends up restricting morality. He argues:

> While the political theory behind deep diversity generates an exaggerated critique of liberalism and a very problematic theory of recognition, the philosophical and moral thought surrounding deep diversity hinges on a moral ontology of modernity that is both unnecessarily restrictive in its account of the moral sources of modernity and lacking any justification of why it is the moral ontology of modernity.[80]

He argues, then, that Taylor retreats from the processes of modernity. Thereby acting to validate the normative content of environmental belonging, tradition and religion. Taylor, it is argued, is too strong in his condemnation of Nietzsche's *non-ontological standard* which some might say would more adequately accommodate the plural dimensions of modern society. This leads Redhead[81] to argue, therefore, that Taylor has missed the potential to ground his ethic in a Nietzschean 'focus on practical forms of agape' which would not leave him 'whistling in the dark'.[82] Redhead claims that Taylor derives justice within the transcendental realm of dream reality and concurs with Bernard Williams that while Taylor 'inhabits planet Earth and relishes its human history, his calculations still leave it being pulled out of orbit by an invisible being'.[83] He continues:

> Like Taylor, Nietzsche views cultural horizons as indispensable to human life. Yet, in contrast to Taylor, a Nietzschean horizon does not denote an unchallengeable moral ontology.[84]

Here Redhead illuminates the relationships between Taylor and Nietzsche which involves the attempt to escape modernity's epistemology as outlined in Chapter 2. Intriguingly, this argument does not delve into Taylor's attempt to solve the commonality-difference conundrum; the processes of change implicit in a fusion of horizons; and the processes of understanding that engage and transform erroneous conceptions of the good.[85]

Another interpretation of Redhead is to argue that Taylor's focus is on the Nietzsche 'of the middle period' which recognises the moral issues implicit in the work of the later Nietzsche. As Abbey has argued in her work on Nietzsche's works such as *Human, All Too Human, Daybreak* and the four books of *The Gay Science*, a different reading about identity and belonging can be found which emphasises engagement and understanding of the factors comprising the modern self.[86] The Nietzsche of the middle period exhibits 'a range of nuanced and delicate analyses, especially those dealing with individual virtues and drives and their multiple manifestations'.[87] At this point in his career, Nietzsche had yet to give way to the extremism and exaggeration of his later work. The later Nietzsche, moreover, emphasised plurality and incommensurability which lives on in the celebration of all that which has been marginalised and submerged.

Redhead ostensibly extends the argument made by Taylor's critics in celebrating the misology of reason while oblivious to the moral dilemmas ignored by the postmodern tendency to embrace dearth, difference and in some cases localism. Critics often ignore the liberal thread within Taylor's care for the self reflected in his recognition of the place of culture, language and morality.[88] In barely acknowledging Taylor's efforts to solve the commonality-difference conundrum Redhead overlooks the pathway developed by Taylor's post-liberalism which embraces difference while transcending superficial differences; this is reflected in his extension of Gadamer's 'fusion of horizons'.[89] As discussed earlier, the search is for a common appreciation of that which we share which disturbs prejudice but embraces the background values of significance subject to principles of reasonableness and toleration. Critics miss Taylor's effort to accommodate difference within that common appreciation and thereby they perpetuate the categorical error that Taylor imposes a theistic vision of the good on communities; leaving less and less room for reasonable divergence over environmental and social differences. These arguments miss the point of Taylor's attempt to provide public space that would allow environmental religious and social beliefs to be given expression and allow different values to flourish.

Perhaps the most fascinating engagement with Taylor's interpretivism is White's argument that Taylor emphasises 'the necessity of maintaining space for "alternative modernities" within which the exploration of external moral sources is central'.[90] White's balanced approach, I argue, can work in conjunction with Taylor's search for fullness such that his poststructuralist ontology of identity/difference does not operate with a sharp, master duality of 'fullness' and 'dearth?'[91] As White argues:

The answer simply is that this duality has no master status. I propose rather a pair of terms intended to highlight what I take to be two crucial aspects of the human condition. I try, of course, to present arguments to support my claims, but the question as to whether this dual perspective is of any real value is ultimately answered only by the quality of the conversations and insights it hopefully fosters.[92]

In supporting Taylor's solution to Nietzsche's conundrum it is again White who observes that 'Taylor stands to the Nietzscheans as his secular critics stand to him'.[93] God might be dead for misologists but this leaves many believers excluded and marginalised.

Here, it is important to remember our earlier discussion that Taylor adapts Heidegger's work on the *clearing* provided by language. For Taylor, it is through language that humanity has emerged from nature but without being cut off from it. The clearing provided through language can be used by humanity to glimpse not only nature's extrinsic values but also its intrinsic value. These values can be 'explained by them [humanity] as something they cause, or one of their properties, or as grounded in them'.[94] The *clearing*, therefore, is related to humanity, but is not ultimately controlled by it. Taylor explains that 'the clearing is the fact of representation; and this only takes place in minds, or in the striving of subjects, or in their uses of various forms of depiction, including language'.[95] Following Heidegger, Taylor argues that the clearing is the space through which expressive language opens up the role of the human person as a 'Shepherd of Being'. The notion of shepherd connects with Heidegger's exploration into the parameters about what it is to be a person (*Dasein*). Moreover, it shapes Taylor's interpretation of Heidegger's supposition that modern philosophy offers predominantly an instrumental stance towards being a human person.[96] Language, therefore, is the conduit through which what is 'significant' for an identity is revealed as a dimension made manifest in the clearing itself, opened to us by our ability to speak.

Following Taylor's work on the difficult Heideggerian notion of the *clearing* it is useful to remember that it is where language acts to disclose humanity's relationships with other entities and the world. Taylor explains that Heidegger argues that the clearing is where the 'mortal and divine beings, earth and sky' are seen to come together. The *clearing* is both *Dasein*-related, but yet not *Dasein*-controlled – where language is not arbitrary and the properties of entities shows up without distortion and where their authenticity is revealed in context. It is through the capacity to utilise language, Taylor argues, that it is possible that we can rebuild our world so that the values of significance are revealed in an open and transparent civil society. Taylor builds on these ideas to construct a political strategy designed to transcend the modernist assumption that language does not count in recognising the intrinsic value which shapes a full life. Taylor's reconstruction of the role of language therefore develops political space to not only respect different values but to combine these values in a democratic manner.[97]

In Heideggerian fashion, Taylor aims to counter the dominant analytical and epistemological perspective in philosophy that assumes that neutrality reflects scientific and precise solutions, but which only ignore people's background values. Perhaps a different way to explain the relevance of Taylor's development of language theory involves his two main principles within his hermeneutic model. His first principle involves recognising that culture, language and tradition shape people's being-in-the-world. His second principle reflects his commitment to the development of the politics of recognition in terms of respect and autonomy. Taylor's model, then, is not hostile to Patrick's[98] critical hermeneutic where the ontological level might, but need not, influence the advocacy of a common good at the political level.[99] That understanding of modernity, as Taylor consistently argues, could lead to a critical analysis of modern practices based on notions of liberal neutrality. Thus, central to Taylor's perspective is:

> The thesis is that our identity is partly shaped by recognition or its absence, often by the misrecognition of others, and so a person or group of people can suffer real damage, real distortion, if the people or society around them mirror back to them a confining or demeaning or contemptible picture of themselves.[100]

Taylor's conception of a reconciled society is not created through an overlapping consensus, nor a mere *modus vivendi*. Rather, it reflects a strong conception of reconciliation which challenges Neo-Nietzschean attempts to problematise or disseminate the structures of modernity while allowing difference free reign, even though the moral implications are rarely given consideration. His alternative approach is to search for that common ground which is based on humility and respect towards others and the natural environment. Reconciliation, moreover, involves a dialectical process of mutual engagement between the ontological and advocacy levels and therefore does not privilege any one particular source of the self. Taylor's approach to politics in this respect is similar to the Nietzsche of the middle period where understanding and moral improvement can be combined with an approach which sharpens discourse and works to reconciling deep divisions in a spirit of common human endeavour.

Taylor's interpretivism and the environment – further reflections

Perhaps one further way to illuminate Taylor's model is to consider the dilemma between individual freedom and environmental restraint. This has been discussed by environmental theorists such as Arias-Maldonado, Dryzek,[101] and Eckersley.[102] Due to the limitations of space I shall focus on Robyn Eckersley who argues that:

> [T]here is more mileage to be gained by enlisting and creatively developing the existing norms, rules, and practices of state governance in ways that make state power more democratically and ecologically accountable than designing a new architecture of global governance *de novo*.[103]

Eckersley's argument is that the state must be involved in guiding the transformation of social consciousness towards ecological values. She has argued that the state must have a role to play in the construction of any critical ecological politics. Extending and exploring Eckersley's important argument involves clarifying how modern societies can be constructed to understand not only differences between reformers and other sustainability advocates, but how they can nurture commonalities to transcend shallow interpretations concerning the value of the natural environment.

While Eckersley has acknowledged the need to re-interpret our understanding of autonomy, there has been little analysis of why people would agree to a green state.[104] This line of thought also reveals a dilemma for interpretivists concerned with creating the conditions that would lead to different and new interactions with the natural environment. Interpretivists argue that the environmental and social conundrums facing modern communities involve not only transforming the state but working to create new world-views. Implicit in my argument is that we need to deepen our appreciation of our place in the natural environment. Interpretivists emphasise our commonalities with the natural environment and this involves transforming our attitude about what it is to be in the natural environment.

Eckersley provides useful ideas for creating green communities and economies without resorting to procedural models. For example: countries such as Australia could (and should) push for the creation of a multilateral fund following the model of the Montreal Protocol.[105] The state might finance the incremental costs of mitigation measures by developing countries. It could also support the idea of voluntary targets for developing countries that provide no sanctions if they underachieve but significant rewards if they are met, and the option of selling their carbon credits if they overachieve.[106] Of course, the state is an important element in creating improved environmental outcomes in the social fabric, but it is also important to develop connections between people and the natural environment.

Essentially, Eckersley offers an imminent critique of critical approaches based on Habermas's discourse ethic and argues that morality and justice are determined in real discursive dialogue where: these 'are not matters that ought to condition or constrain the rules of discourse in advance of the dialogue'.[107] I argue that Eckersley's important line of thinking can actually work together with Taylor's search for new world-views such that people are not separate from the contexts in which they live. Accordingly we need to re-theorise not only the way we manage the global commons but also how we understand the role of language.[108] This task is undertaken by examining the social imaginaries that lie behind our discourse and narratives that guide our social collectives and consciousness. It is these social forces and vectors that require deeper articulation and analysis.

8.6 Conclusion

This chapter has argued that Taylor tackles the limitations of Enlightenment reason and modernity's preoccupation with procedure. He emphasises the art of interpretation as a means to provide compelling reasons why people should search for what is common despite their differences. While offering a commitment to the politics of the common good, he invites us to think about why some people are required to give up their beliefs if they are to secure the benefits associated with life in a liberal democracy. It is therefore problematic to claim that Taylor's post-liberal inspired interpretivism imposes a common good as the means to solve these normative moral issues. It has also been argued that Taylor's work on the possibilities of the common good reflects a complex combination of ideas and political manoeuvring in solving the commonality-difference conundrum. Notwithstanding the attempts to construct virtual republics, there has been little response to the communitarian definition about what it means to be a person. To be a person also involves collective social infrastructure to solve ecological and social problems that reflect common purposes.

Central to a Taylorian framework is his engagements with the ideas of Jürgen Habermas. It has been argued that Taylor not only re-evaluates the work of both Herder and Humbolt concerning language and 'representational or linguistic consciousness' but also explains how a universal mechanism lies within a particular linguistic phenomenon. Taylor continues that the central difference between these two traditions of thought concerns the role that language plays within the art of interpretation. As Taylor has stated, this is 'the central question' which this chapter has explored. The interpretations of Taylor's work in this chapter parallel the interpretation of Stephen K. White who observes that Taylor opposes the Neo-Nietzschean traditions, as secular theorists oppose his theism. Then the chapter proceeded to argue that the public sphere must be open to all significant values as a means to accommodate these differences in a political mosaic based on justice and fairness. Taylor not only articulates life-goods but also works to overcome the relativism that 'lies within the postmoderns' suspicion of the causal power of our culture's goods'.[109]

Therefore, Taylor's argument involves an exploration into the realist assumption that conflict between people is the variable left unanalysed in democratic and liberal frameworks. This argument is based on his Hegelian observation that tolerance and respect should not be used to obfuscate the real differences which exist between people, communities and societies. Only when these elements are considered is it possible to fully begin to recognise the unique role of Taylor's thinking in reconstructing communities that develop people's freedom with a spirit of recognition that acknowledges people's being-in-the-world. Further, it must be remembered that Taylor is not offering an ethic which legitimates cultural insularity and particularist conceptions of the good. Rather, his argument is that a society that accommodates strong evaluations is likely to accommodate difference with justice than do those frameworks which simply rank differences according to some external metric.

9
Critical Perspectives: The Taylor–Rorty Debate

This chapter engages with environmental matters by using ideas from Johnathan Marks, Richard Rorty and Taylor. I deepen the discussion of the role of language to explore processes of equal facilitation in the context of environmental values. Through face-to-face community relationships, members of communities can be better informed about wasteful and damaging resource consumption and the ecological burdens to be borne by all.

9.1 Introduction: environmental values

Since the seventeenth-century Enlightenment, the pre-eminent economic and political theories have offered a controlling, dominating and technical attitude towards the natural environment. According to political interpretivists such as Dreyfus, Gadamer and Taylor, these theories have been inspired by an economic vision of the value in the natural environment. The end result of this instrumental process has been a disenchanted conception of the natural environment – a world that is devoid of all intrinsic and theistic meaning. In this chapter I develop the interpretive work of Dreyfus, Gadamer, Nussbaum and Taylor to explore the natural environment as a shared ecological and social commonality.

Further I focus on the supposition that the natural environment possesses meaning and that new political structures are needed. I explore how Taylor's work might allow us to engage with multiple cultures concerning matters at the heart of ecological politics by using his work on Richard Rorty. Central to the argument is the belief that interpretivists offer processes of equal facilitation and maximisation that can be used to create awareness of the significance of the natural environment.

This approach differs from theorists such as Rorty who has debated with Taylor on these issues. The chapter develops the Rorty-Taylor debate to understand the role of practical reasoning and how humanity interacts with the natural environment.[1]

A central theme to keep in mind is that new ways to govern modern communities are needed if we are to align our activities with the ecosystems and environments in which we live. In this chapter I aim to deepen our analysis of the relationships between humanity and the natural environment by utilising the resources associated with the political art of interpretation. I develop earlier discussions by using interpretivist ideas to comment on the environmental differences between interpretivism and Rorty's important work which discusses the natural environment. Taylor's work provides a multifaceted understanding of the specificity of communal life, and the common purposes that exist for everyone as part of the natural environment. The chapter has seven key sections following this introduction. My first task in Section 9.2 outlines some criticisms from Johnathan Marks. Section 9.3 explores the notion that we are voices from nature by introducing the work of Baudrillard and Rorty. Section 9.4 deepens these critiques using Rorty's work on Dreyfus and Taylor. Section 9.5 outlines the role of language in the Rorty-Taylor debate. Section 9.6 outlines the space opened up in language. Section 9.7 develops interpretivism to search for intrinsic value. In Section 9.8 I conclude the chapter by keeping in mind how interpretivists attempt to escape the iron cage of bureaucratic and instrumental reason. The key theme in this chapter is the Taylorian idea mindful of the need to explore connections between our concepts, categories, intuitions, perceptions and nature.

9.2   The voice of nature: further objections and contrary opinions

A useful way to explore the environmental implications of the interpretivist approach is to review some important criticisms from postmodern quarters. One obvious criticism is that interpretivists such as Dreyfus, Gadamer, Heidegger, Nussbaum and Taylor are not generally regarded as environmental philosophers. This objection can be addressed when interpretivists explain that the full implications of our actions in the world cannot be explained simply by economic and scientific methods. Of course, instrumental theorists and technocrats respond by arguing that environmental problems can be solved using the techniques of sustainable management. However, questions concerning the type of world we are creating remains open. Modern social science ignores the interpretive approach and in doing so detaches people from their internal sources of value that express and perceive their place in the world.

A further problem with current instrumental horizons of meaning is reflected in patterns of sustainable development: that is, instrumentalists do not question how modern society has come to its present anthropocentric stance to the world. This is because sustainability remains defined within an economic growth paradigm, itself reflecting a limited ontology.[2] The ontology on which sustainability is based does not consider fully how sustainable patterns will be achieved in the current system; that is, it relies on anthropocentric ecological policy where environmental restraint is shown as necessary for human purposes only. In an important essay Jonathan Marks criticises Taylor for elevating community obligations over individual autonomy.[3] He argues that Taylor relies too heavily on Rousseau's physiodicy which is said to fail to develop fair and just solutions. Marks claims that Taylor's emphasis on community and significance[4] reflects a mis-reading of Rousseau's thinking. Marks argues:

> But if Rousseau's understanding of obligation to community is too exclusionary, it may point out one sense in which Taylor has too inclusive an understanding of such obligation. For if the community earns a citizen's loyalty by providing him with an identity rather than moral citizenship, with moral depth rather than morality, it is difficult to see how one avoids obligations to very bad communities.[5]

Marks assumes that identity and moral depth leads to a politics that justifies a volk or community-centred focus. This argument, was discussed earlier when we analysed Habermas's work who claims that Taylor's politics of species preservation would lead to the elevation of environmental interests over those of people. That is, rather than create fair principles of justice the Taylorian argument replaces principles of justice with arguments to preserve the natural environment.[6] This implies that Taylor is using our relationships with the natural environment to determine what is right or wrong. That is, he is operating with an ethic that elevates spiritual and environmental values while gliding over ethical and moral theory.

However I use Taylor's argument that the natural environment is a source of the self that resonates within people when they come in contact with it.[7] That is, something important is lost when our lives are directed by economic forces and instrumental social imaginaries. Both Habermas and Marks do not explore how Taylor attempts to create a society that recognises the importance of common and collective goods that would nurture these feelings and values. His ethic does not use nature or community as external standards of justice – justice in Taylor's view involves two stages. First, justice is about understanding and engaging with the other on fair terms of recognition. The second stage involves entering into dialogue and understanding the perspectives of other interlocutors to work out political principles of justice that allow people and communities to live together. It is therefore a categorical error to focus on community without consideration of these other elements in the interpretive strategy. Chapter 3 and 4 showed that Taylor develops 'supersession arguments' which involve:

> (W)here we show that there is a rational path from A to B, but not in the reverse direction. But we have to see this path not only as a line of argument, and not only as an actual transition, but as both together.[8]

Here both ethics and morality are combined in this thinking such that the path to a dialogic and just society involves providing a best-account argument. The path developed by Taylor does not subscribe to the voice of nature, but explores how we are in touch with the world. As self-interpreting beings we must explore how we are a part of nature.

Often critics ignore not only the interpretive dimensions, but also the optimistic use of practical reasoning that explores the role of perception and belief. These steps open people to the possibilities that we are in contact with the universe. For these reasons the environmental issue reflects Taylor's ontological arguments opening citizens to a fusion of horizons. This involves a moral sense which was discussed earlier where each individual is opened to different goods and values in a quest to create better societies. Indeed, Rousseau coined the phrase the 'voice of nature', which has been criticised as part of a complex anti-liberal *modus vivendi.* Marks, however, argues that interpretivists might unwittingly end up justifying very bad communities,[9] but this argument ignores the philosophical journey involved in understanding our place in the world. The interpretivist thinking is that it is important to explore how our moral sense connects us with the world, where moral sense is a way of thinking which recognises the values important for each individual. Marks' argument implies that the ethical formation advocated by interpretivists inevitably supports a cultural and uniform political structure.

This is because the common good could be used by political administrators who have the mandate to override individual autonomy. For Marks, Taylor's support for community and identity conceivably rides roughshod over individual autonomy. This problem has also been expressed by Richard Wolin in his work – *Heidegger's Children*[10]– when he expressed reservations with those thinkers reliant on Heideggerian politics. This concern applies to Taylor as he is heavily influenced by Heidegger's political thought, condemned by many for its affiliations with National Socialism. The environmental issue that Taylor develops using ideas from Heidegger and Rousseau, however, is an ontological consideration and at that level Taylor concurs with aspects of Deep Ecology.[11] He differs in his enthusiastic celebration and recognition of theism.[12] A trite response is that a theorist of Taylor's calibre would certainly not endorse the illiberal politics associated with Deep Ecology, right wing Heideggerianism or Nazism. He is well aware of the dangers in Heideggerian visions of a technological world. Of course, he is not an earth-based environmentalist and has stated that his environmental perspective emanates from the interplay between the politics of recognition and the role of theism in a secular world.[13] His work on theism and the sources of the modern self reflects an enduring commitment to fundamental human rights and the need to recognise other values.

9.3 The voice of nature: Baudrillard and Rorty's critique

From what might be referred to as postmodern quarters, Baudrillard and Rorty claim that Taylor's moral realism supports intrinsic value arguments but never proves them. As a result, important philosophical concepts are left hanging in the air because Taylor blames anthropocentricism for the spiritual decline in modern societies. My argument is that he uses the work of Gadamer and Merleau-Ponty to guide an enthusiastic commitment to practical reason; namely, a commitment that respects difference in a good society. The difference-commonality conundrum is addressed in a political structure that works to develop commonalities between citizens and respects difference.

Accordingly, the diverse elements making up a community are accorded respect in a political and social structure nurturing such commonalities. This, of course, is designed to bring about respect and acknowledge different views over the natural environment. Taylor advocates a politics that focuses on commonalities, and he explores this theme in his work on federal structures.[14] For Taylor, a federal structure is like a patchwork quilt that when combined creates a meaningful whole. That is, when the elements of the structure are assembled together the outcome is a political mosaic which when combined forms a greater whole. However, critics allege that political problems emerge if we reconcile ourselves with the natural environment. For example, Baudrillard argues:

> In short, it is not by expurgating evil that we liberate good. Worse, by liberating good, we also liberate evil. And this is only right: it is the rule of the symbolic game. It is the inseparability of good and evil which constitutes our true equilibrium, our true balance. We ought not to entertain the illusion that we might separate the two, that we might cultivate good and happiness in a pure state and expel evil and sorrow as wastes. That is the terroristic dream of the transparency of good, which very quickly ends in its opposite, the transparency of evil. We must not reconcile ourselves with nature.[15]

Baudrillard's perspective assumes that communitarian interpretation leads necessarily to a disciplined and uniform political structure. This position seems to reflect a misology towards reason and rationality. While such thinking locates and explores the sources of discordant power, interpretation is required to imagine a viable response. Baudrillard's thinking seems to celebrate an anti-realist epistemology that cuts us off from nature. From the ideas presented by Baudrillard, Marks and Rorty the challenge is to explain what Taylor means when he uses Rousseau's argument that humanity is the 'voice of nature'.[16]

9.4   Rorty's critique of Dreyfus and Taylor: the philosophy of nature

To this point it has been argued that interpretivists utilise hermeneutic strategies as they concern morality, interpretation, the theorisation of politics and the art of understanding knowledge. In terms of ecological politics, these diverse strands of thought require careful attention and development. Taylor argues for new insights into the human condition and states that human striving for meaning and understanding is needed. That is, citizens and communities do not even begin to see where we have to restructure our societies as long as we 'accept the complacent myth that people like us, enlightened secularists or believers, are not part of the problem'.[17]

As we shall see in the sections that follow, Rorty and Dreyfus and Taylor differ on the role language and perception play in the formation of knowledge. Rorty's approach challenges the interpretivist strategy of

Dreyfus and Taylor; he invites interpretivists to prove that language is part of a process that reveals the essence of an object, or thing. This is similar to a charge levelled against environmental ethicists in general when it is asked how can the interests of natural entities be given expression in law courts, parliaments or other community institutions? In Rorty's view, philosophy can do no more than assemble ideas and suggest possibilities. Rorty concludes:

> [T]here is no authority outside of convenience for human purposes that can be appealed to in order to legitimize the use of a vocabulary. We have no duties to anything nonhuman.[18]

Similarly, Rorty asks how does language put us in touch with the external world and therefore its meaning and intrinsic value. He challenges interpretivists to explain how values external to the human valuing subject can motivate an individual agent. Rorty then urges philosophers to give up on the search for such essential meaning and just focus on the human predicament.

Rorty's criticism asks what Taylor means when he states that humanity is the 'voice of nature' and how these views are shaped by the processes of instrumental reason and secularism.[19] Rorty is representative of that line of critical thinking when he argues that Taylor's concerns are unfocused and offer little in the way of alternatives to modernity's twin dangers: domination and control over others and the natural environment. Rorty's argument implies that interpretivists are too close to the Deep Ecology's axiom that nature possesses intrinsic value. He claims that interpretivists do not offer worked out alternatives to economic and technical solutions; thus, the problems with an anthropocentric culture are unlikely to be solved by the art of interpretation and moral theory. However before we explore the intricate differences between Rorty and Dreyfus and Taylor it is useful to explore the issues that they debate. They are:

1. That the methods of the natural sciences cannot be transferred to the social sciences and important values for a person's identity are dependent on the various contexts in which people operate.
2. That substantive values and contexts are essential to human existence which involves understanding the assertions and arguments presented in discourse. That is, in order to understand moral and political concepts one has to recognise the substantive evaluative terms used.
3. That language is central to understanding the human condition.

These assumptions lead us to the view that language is a means to reveal substantive or evaluative terms that shape moral or political concepts. All three philosophers agree that it is through language that different interpretations are expressed and articulated.[20]

Implicit in Rorty's critique is the argument that humanity is not a 'voice of nature', but a voice from it. Essentially operating from an anti-realist position, Rorty maintains that it is not possible to visualise the world beyond our current concepts and categories. The key issue involves whether the art of interpretation is able to create and develop commonalities such as those between humanity and nature. Taylor strategises the morality, interpretation, theorisation of politics and the art of understanding knowledge. In terms of ecological politics and understanding the world, these diverse strands of thought require careful attention and development in how we construct our communities.

Interpretivists such as Taylor argue that Western societies will pay a high environmental price if we allow this kind of closed world thinking to prevail. That price may be reflected in the form of anomie, cultural disharmony and environmental damage – interpretivists aim to rethink our relationships with the natural environment. They offer a realism that is said to put people in touch with the external world. It is important to remember that these arguments are disputed by Richard Rorty who criticises the politics of reconciliation, the art of interpretation and unproblematic realism. This long-running dispute concerns the distinctions between the natural and social sciences as they impact on how we are to understand our place in the natural environment. Rorty believes that once these distinctions are broken down then the role for philosophy is complete. Dreyfus and Taylor respond to this point by arguing that the methods of the social sciences require new criteria and new thinking to understand humanity's place in the world. They question the potential relativism within Rorty's work and offer a vision of a reconciled society between people and with nature.

In Rorty's view our research efforts should just challenge the normalising discourses of modernity because there is no truth, or objectivity, independent of the rules of discourse constructed by people. All that philosophy can do is to continue the conversation of humanity and suggest some interesting reminders. For Rorty it is impossible to prove the existence of intrinsic value in the natural environment and be motivated by an unknowable thing-in-itself. This argument does not lead to a full rejection of environmental and intrinsic value in and of itself. Rather, it means that no faculty exists for humanity to glimpse the otherness of nature's value. There is nothing outside the text, nor is there anything independent of our human agency on such a view. No deeper sense of reality exists independent of our current practices.

9.5 Rorty's critique: the role of language

Rorty follows Donald Davidson's work on language and the world. Rorty argues that once the distinctions between the natural and social sciences have been demolished the best we can do is to suggest interesting possibilities and reminders. That is, we must recognise that justification relies on what we already accept – the only way to justify a belief is through another belief. For Rorty there is no independent means to get outside of our beliefs and language.[21] Therefore, intrinsic values do not exist. Rorty states that:

> [P]hilosophical reflection which does not attempt a radical criticism of contemporary culture, does not attempt to refound or remotivate it, but simply assembles reminders and suggests some interesting possibilities.[22]

According to Rorty, once the distinctions between the natural and social sciences have been proven to be incapable of providing universal claims to truth then all subsequent evaluative moves prove elusive. Rorty continues in his *locus classicus* – *Philosophy and the Mirror of Nature* – that the differences between the social and natural sciences imply that no external values can command our allegiance. Any realist epistemology must reflect a particular interpretation, and this is because such knowledge reflects a particular horizon of meaning. He concludes:

> Epistemological behaviourism (which might be called simply 'pragmatism', were this term not a bit overladen)...it is the claim that philosophy will have no more to offer than common sense (supplemented by biology, history, etc.) about knowledge and truth. The question is not whether necessary and sufficient behavioral conditions for 'S knows that p' can be offered; no one any longer dreams they can. Nor is the question whether such conditions can be offered for 'S sees that p', or 'It looks to S as if p', or 'S is having the thought that p'. To be behaviorist in the large sense in which Sellars and Quine are behaviourist is not to offer reductionist analyses, but to refuse to attempt a certain sort of explanation: the sort of explanation which not only interposes such a notion as 'acquaintance with meanings' or 'acquaintance with sensory appearances' between the impact of the

environment on human beings and their reports about it, but uses such notions to explain the reliability of such reports.[23]

Rorty continues with the argument that his 'behaviorism' is designed to explain 'rationality and epistemic authority by reference to what society lets us say, rather than the latter by the former'.[24] And so we find the essence of 'epistemological behaviorism' is 'an attitude common to Dewey and Wittgenstein'.[25] This is based on Rorty's view that notions such as 'acquaintance with meanings' offer means to access our world. I argue that this strategy has the potential to return us to closed world thinking without offering a way out of the mind–world dichotomy. Furthermore that this strategy has the potential to return us to procedural approaches without offering a way out of the mind–world dichotomy. Rorty's critique of traditional epistemology, in this regard, involves the claim that interpretation is simply a reflection of a set of beliefs: on this view, it is impossible for us to reveal nature's meaning or our place in the universe. Ultimate meaning and nature's intrinsic value are seemingly beyond our comprehension and this leads to his support for pragmatism.

For Rorty, moral realists such as Taylor cannot prove the existence of external values and all we can do is to continue the conversation of humanity. Accordingly, all knowledge is mediated through the mind and any attempt to understand intrinsic value ends up in some form of essentialism or mysticism. On his view, essentialism is shaped by a subjective and particular discourse that operates at that moment in time. Rorty continues that it is interpretivists such as Taylor who remain enthralled by Cartesian dualisms offering little in the way of philosophical insight. An obvious response is that people actually do experience their being as perceptually in touch with the universe which was discussed earlier in Chapter 5 and developed in the following section of this chapter.

Rorty continues in his view that the moral life is simply a series of compromises.[26] But environmental and social questions do not just revolve around whether Aristotle or Dewey held such views – rather, the question involves whether people are perceptually in touch with the world. Rorty however remains unconvinced that we can differentiate between the truth about astrophysics and how things are in themselves. Rorty continues:

[D]ebates about astrophysics, how to read Rilke, the desirability of hypergoods, which movie to go to, and what kind of ice cream tastes best are, in this respect, on a par.[27]

Rorty then urges withdrawal from ontological commitments and in turn would reject arguments that nature is a source of the self. All statements in turn reflect intuitions and beliefs because there are no essences, accordingly we have no access to the universe, or to intrinsic value. Accordingly, agreements between people are all that is possible which amounts to a deflationary realism where there is no philosophical way for people to stand outside of their practices. In sum, Rorty's view places Taylor as an essentialist who cannot escape from the epistemological cage of modernity.

9.6   The role of language – Dreyfus and Taylor respond

Taylor responds to the arguments made by Rorty using his 'best account' political strategy. Taylor argues that the distinctions between the social and natural sciences have not been demolished. He challenges Rorty's approach by demonstrating that people have pre-conceptual coping skills which impinge on the world.[28] Of course, interpretivists are not arguing that we are in touch with the universe in and of itself in the way a scientist would encounter the world. Rather, Dreyfus/Taylor argue that as engaged agents people are in touch with the universe in their everyday lives. Rorty, however, maintains that Taylor's supersession argument[29] is just another attempt to explain how things are in themselves. Dreyfus argues that Rorty's approach misses the need for interpretation. Dreyfus explains:

> But if, as in Rorty's projection, all objective truth were settled, and there is no other area of serious investigation of shared phenomena, abnormal discourse could only be the expression of an individual's subjective attitude towards the facts. And once this meta-truth was understood, there would be no place for hermeneutic efforts at commensuration. Indeed, there would be no sense of translating one discourse into another by trying to make them maximally agree on what was true and what false as proposed by Gadamer and Quine, since all discourses would already agree on the objective facts. All that abnormal discourse would amount to would be the expression of private fantasies, and resulting pro and con attitudes towards the facts. And all that would be left in the place of Rorty's kind of hermeneutics would be the Derridean notion of the play of discourse about discourse.[30]

Rorty's stance to politics relies on irony and compromise. But Dreyfus and Taylor in their respective approaches claim that Rorty misses the point of ontological reflection and our pre-conceptual abilities which open us to nature. Here, Taylor responds:

[The] understanding which is relevant to the sciences of man is something more than this implicit grasp on things. It is related rather to the kind of understanding we invoke in personal relations when we say, for instance: 'I find him hard to understand'; or 'at last I understand her'. To switch for a minute to another language, it is the kind of understanding one invokes in French when one says 'maintenant on s'entend'. We are talking here in these cases of what you could call human understanding, understanding what makes someone tick, or how he feels or acts as a human being.[31]

For Taylor, once an action has been disclosed it reveals itself, its purpose, as well as the practices in which it operates. Irrespective of how an action is interpreted, it reveals itself as significant for its 'being-in-the-world'.[32] This means that all discourse can be settled in that all values are no more than a reflection of subjective preferences – all that remains is relativism and the play of difference. Here, Dreyfus argues that Rorty's search for 'consensus' and 'agreement' ignores the background values shaping language and interpretation. Dreyfus notes:

They hope that by seeking a shared agreement on what is relevant and by developing shared skills of observation, etc., the background practices of the social scientist can be taken for granted and ignored the way the background is ignored in natural science.[33]

Dreyfus's 'strong hermeneutic' involves embedded coping that is characteristic of humanity's perceptual skills that can be confirmed and checked against reality.[34] People are directly in touch with the world and its significant values. Modern secular theories fail to recognise this and have put humanity on a wrong pathway. Last, Dreyfus and Taylor argue that Rorty ignores how better interpretations build up our appreciation of the world. Dreyfus underlines this problem in Rorty's strategy:

As long as someone feels that something has been left out of the objective account he may be inclined to propose and defend an interpretation of what that nontheoretical residue is and what it means for human action.[35]

Thus, Rorty's ironical stance finds itself immersed in relativism and does not critically interpret the practices that shape our lives.[36] Rorty's refusal to examine how the social sciences provide different ways to understand our place in the world involves not only thoughts and ideas but also skilful coping and our everyday ability to live in the world. While it might be the case that 'science is in fact zeroing in on (one aspect of) the physical universe as it is in itself' it does not explore how our experiences influence our actions.[37] Taylor's developed argument is that:

> Once we see the emptiness of the myth of the Given, our problem is somehow to bring this free spontaneity together with constraint. In order to stop the oscillation between the need for grounding which generates the myth of the Given, and the debunking of this myth, which leaves us with the need unfulfilled, 'we need to recognize that experiences themselves are states or occurrences that inextricably combine receptivity and spontaneity',[38] we have to be able 'speak of experience as openness to the lay-out of reality. Experience enables the lay-out of reality itself to exert a rational influence on what a subject thinks.[39]

Our experiences themselves put us in touch with the world: putting the point simply we do not always utilise our thinking skills when acting in the world, as I argued in Chapter 5. Our individual beliefs do not simply mirror the in-itself-of-the-world, but involve our active engagements in that world. Ultimately, Rorty moves too fast and ignores our embedded grasp of reality as it gives us access to the world. Rorty continues that Taylor sees a relationship with the world more primordial than representation but in fact there is no necessary 'break between nonlinguistic and linguistic interactions for organisms (or machines) with the world'.[40]

However according to Taylor it is Rorty who does not explore how the 'sciences of man' have different approaches through language to reveal meaning in the world. Rather, Taylor's point is that there are various new insights involving our interactions with the world.

- First, Taylor argues that the philosophical task should be to reflect on existing social and environmental practices by exploring how our interpretations explain our engagement with the world.
- Second, in following Taylor it is important to discuss the concept of the 'world' using the power of language to open up new ways to think about our world.

For Taylor, language provides a space where things are revealed and this has implications for environmental ethics. Language is a means through which people express the values of significance in their lives. Gadamer explains:

> [L]anguage in which something comes to speak is not a possession at the disposal of one or the other of the interlocutors. Every conversation presupposes a common language, or better, creates a common language. Something is placed in the center, as the Greeks say, which the partners in dialogue both share, and concerning which they can exchange ideas with one another. Hence reaching an understanding on the subject matter of a conversation necessarily means that a common language must be first worked out in the conversation. This is not an external matter of simply adjusting our tools; nor is it even right to say that the partners adapt themselves to one another but, rather, in a successful conversation they both come under the influence of the truth of the object and are thus bound to one another in a new community. To reach an understanding in a dialogue is not merely a matter of putting oneself forward and successfully asserting one's own point of view, but being transformed into a communion in which we do not remain what we were.[41]

This passage is motivated by a philosophical idealism often misconstrued by political ironists such as imposing commonalities on people. From Rorty's perspective it seems that language can be used as a vehicle by cunning political actors and its seductive powers must be resisted. Language, in Rorty's view, does not disclose anything with indubitable certainty.[42]

Interpretivists respond, arguing that the power of language provides a space through which people arrive at agreements and understanding. The space opened up by language connects with the values revealed in skilful coping (perception). Chapter 2 shows that this thinking concerns how disclosing beings are directly in touch with the world. It is not simply a matter of asserting the existence of intrinsic value, but involves the revealing power of a being capable of reflective thought. In Taylor's view, truth is not present as a statement, but as an experience of an event, more precisely as the experience of strife. The beings are susceptible to being questioned with respect to their truth precisely because they stand on their own. As Ross argues:

> They stand on their own thanks to a preserving 'agent' that stands as their 'ground', what Heidegger calls the 'earth'. And it is this relation between world and earth, understood as a conflict between the

appearing and opening of beings (world) and withdrawing and sheltering of beings (earth), that sets Heidegger on the path of thinking what he terms the 'double concealing' of truth.[43]

The question that follows immediately from Ross's observation is whether language is simply a series of 'marks and noises' as Rorty claims or whether language is part of an interpretivist process which discloses entities and objects. I use this debate and metaphor in order to explain how it is through language that we provide richer interpretations of entities, values and things.

Here I use Taylor's work on Heidegger's famous example of the disclosure of the purpose of a hammer. The example is designed to reveal that our interpretations allow us to tap into reality. This metaphor is useful when exploring Taylor's interpretation of language to reveal meaning, reality and value. Taylor refers us to Heidegger's argument that:

The hammer possessed in fore-having is ready-to-hand as a tool. When this becomes an object of an assertion, with the very construction of the assertion a shift occurs in the fore-having. The ready-to-hand tendency to think 'with what?' becomes the 'about what?' of a referential assertion. The view of the object in preunderstanding is now focused on what is merely 'on hand' in the ready-to-hand ... And readiness-to-hand as such passes into concealment.[44]

Here disclosing the hammer as an object is at the expense of concealing its existence as a tool. In a similar manner, it is possible that disclosing the intrinsic value of nature reveals the otherness of nature which cannot be reduced to apparatus or technique.

According to the interpretation provided by Taylor it is only in the act of hammering that the object becomes the entity we refer to as a tool – a hammer. Prior to this act, our understanding of it is limited to its appearance as a shape rather than as an object within a context that gives the shape a form or meaning. Here Taylor is explaining some important distinctions between how knowledge is developed. From one angle, knowledge is shaped by the context it arises to us. Another angle reveals our more primordial pre-understanding of entities and things which are disclosed by how we interpret them.

For Taylor our pre-understandings of meaning and value come to light through our interpretations of them. Rorty disputes Taylor's argument and maintains that the metaphor of the hammer has meaning as a tool in a context of use that passes through a series of interpretations that are subject related. In contrast, Taylor appeals to the universal characteristic of the hermeneutic method, which maintains that, irrespective of our interpretations once the act of 'hammering' is performed it discloses its existence at that moment in time. As Taylor explains:

> We aren't deriving this from the nature of 'rational animal'. It is, on the contrary, purely derived from the way of being of the clearing, by being attentive to the way that language opens up a clearing. When we can bring this undistortedly to light, we see that it is not something we accomplish. It is not an artefact of ours, our '*Geachte*'.[45]

For Taylor, there is an indubitable sense in which things can be disclosed thereby revealing the different aspects of their being. In a similar way, the intrinsic value of nature can be revealed as part of us; that is, the otherness of nature is not only part of us but can be expressed by us.

9.7   Developing interpretivism and the search for intrinsic value

This section explores how Taylor moves beyond Rorty's closed worldview that we have no duties to anything nonhuman. Here Taylor refers us to Hans Georg Gadamer's work on moral awareness as a means to explore the value in nature. Clearly, this thinking is relevant to our discussion concerning the connections between language and meaning in the world. Gadamer argues that agents imbued with a moral sense have a means to glimpse the intrinsic value of the natural environment. However, Gadamer argued that:

> That nature is beautiful arouses interest only in someone who 'already set his interest deep in the morally good'. The interest in natural beauty is akin to the moral' which resonates with our wholly disinterested pleasure – i.e., the wonderful purposiveness (ZweckmaBigkeit) of nature for us, it points to us as the ultimate purpose of creation, to the 'moral side of our being'.[46]

Piercing the obscurity of Gadamer's moral dimension involves working through connections between meaning, morals and intrinsic value. Following Lars Samuelsson I am of the view that Gadamer's moral dimension reflects normative non-natural reasons that are not reducible to other reasons.[47] Samuelson posits a connection between normative non-natural reasons and intrinsic value which can be interpreted as the moral dimension. He provides perhaps the clearest argument that allows us to understand what Gadamer and Taylor mean by the moral dimension putting us in touch with the intrinsic value of the natural environment.

Samuelsson argues that no naturalist conception of value can be totally correct. He argues that this is because the normative is not reducible to the natural and 'any identification of value properties with natural properties is bound to fail'.[48] Therefore, intrinsic value is a non-natural property and intrinsic value is plausibly a mind-independent property. I argue that the non-natural dimension reflects the moral view held by Gadamer and Taylor. That is, it is through the moral dimension that we encounter experiences of mind-independent intrinsic values. Here Samuelson argues therefore that we are justified in believing in the existence of such values, and this argument connects with Taylor's work on interpretivism and language.[49]

Gadamer and Samuelson assist to explain how the moral dimension provides a critical perspective on the supposition that nature is an economic and technical good. The dominant economic and procedural approach sets in place a cultural malaise based on not only instrumental reason and perpetuates closed world thinking. Taylor argues that to ignore how our moral sources involves our perceptive abilities is likely to continue to detach people from the natural world. Arguably, closed world thinking leads to a nihilistic vision of the world – a world devoid of intrinsic meaning and virtue. Here Taylor's work on language is relevant in offering a perspective that challenges the closed world thinking implicit in those views which claim that the natural environment is a simple social construct. Communitarians and interpretivists counter by arguing that it is human beings who have the moral frameworks that allow an appreciation of nature's value to emerge.

Furthermore, it is through language that people are able to express the values that we feel when we perceive and interact in the world. There is a sense in which nature possesses intrinsic value and it is for embodied agents to express it. This thinking builds a bridge between humanity and nature where different features of nature's value are revealed to beings with a moral sense. Of course, Gadamer and Taylor are not arguing that perception trumps the power of reason such that the natural environment acts as normative ideals. According to Taylor, therefore it is Rorty's deflationary realism which is limited by two further key problems from Gadamer's perspective. First, Rorty does not fully offer a way to stand outside of our practices to criticise the institutions of modernity, which require improved interpretations that build up richer pictures of the world. Second, Rorty's self-proclaimed epistemological behaviourism does not fully explain how the conditions for agreement might be brought into play through language and discourse. On this point, Taylor returns to his work on the differences between the natural and social sciences. For these reasons, Taylor has argued that we are part of nature which involves different ways of orienting and thinking about it. Rorty continues to criticise this argument:

> Realism becomes interesting only when we supplement plain speech and common sense with the 'in itself' versus 'to us' distinction. Taylor thinks that this latter distinction cannot simply be walked away from but must be dealt with. I think neither he nor anyone else has explained why we cannot just walk away from it. Such an explanation would have to tell us more than we have ever before been told about what good distinction is supposed to do. I keep hoping that Taylor, as fervent an anti-Cartesian as I, will join with me in abandoning it.[50]

Taylor notes that it is important to understand the distinction between 'in-itself' and 'for us' if we are to provide richer interpretations of our place in the world.[51] According to Taylor it is Rorty who narrows the role of interpretation and disclosure and remains in a Cartesian worldview whereby we continue to fail to engage adequately with the natural environment. And so we remain in closed world structures which reflect Rorty's tendency to foreclose further debate and interpretation while ignoring direct coping, interpretation and the best-account argument, which is a key component in exploring how we are in touch with the world.[52]

Here Taylor argues that it might be the case that science is zeroing in on one aspect of reality but there are other features of our reality which involve how ' [visions] of reality [impact on what is] liveable'.[53] According to Taylor it is Rorty's interpretation of the disclosing power of language which can lead to a situation where one becomes 'incapable of saying important things, or forced to banalise about important things'.[54] However, the task of philosophy is to assist in understanding the world in which we live. It is for this reason that Taylor's work offers a method of interpretation through which both sides of a dispute can be understood. Taylor argues that:

> [T]hose writers are correct who maintain that scientific knowledge has to do with our learning to grasp and identify the kinds of things there

are and their causal action. The 'realist' view, defended by writers like Harré, is that we explain phenomena by showing them to flow from the operation of the kinds of thing that there are, where our conception of such kinds cannot be reductively eliminated in favor of some type of concept.[55]

Here Taylor explains how different aspects of reality are revealed. A type of science might reveal how things are in themselves but that is only one way of disclosing them. For the purposes of my environmental argument, the meaning and intrinsic value of nature flow from the type of entities that they are.[56] On this issue it is again Dreyfus who interprets Heidegger's work as providing a means to reveal different features of reality. He quotes Heidegger that 'of course the statements of physics are correct. By means of them, science represents something real, by which it is objectively controlled. But science always encounters only what *its* kind of representation has admitted beforehand as an object possible for science.'[57] Dreyfus reminds us that Heidegger is revealing different facets of reality using different techniques and frameworks. No unique language exists that correctly describes reality as it is; however, there may be many languages each of 'which correctly describe a different aspect of reality'.[58]

9.8   Conclusion: an interpretive and environmentally aware civil society

Interpretivists offer a political strategy to safeguard the natural environment and future generations. This is to be achieved through face-to-face community relationships in a democratic structure that is attentive to the otherness of nature. A concern for such values and the future of the world, however, can be curtailed within the proceduralism of modern liberalism and the postmodern stress on difference. A new, interpretive-inspired age might be one where people relate to, and cope with, the turbulent environment in which they live and act.

Interpretivists suggest that there is a connection between intrinsic value in the world and how external values might put people in touch with the natural environment. Postmodern contingency cuts us off from our expressive sources of the self precisely when interpretation is most needed. From this it follows that it is possible that our coping skills orientate us to the world, where language is the conduit through which the otherness of nature is given expression. That is, our explorations into meaning, nature and steady-state social arrangements provide a further challenge to capitalism and technological modernity. In this, I argue that interpretivism connects with the environmental precautionary principle which states that 'where there are threats of serious or irreversible damage, lack of full scientific certainty shall not be used as a reason for postponing cost-effective measures to prevent environmental damage'.[59]

These arguments reflect two interlocking premises. First, I explore the question of why modern secular societies have had difficulty recognising the otherness of the natural environment.[60] Second, I argue that direct moral realism is useful when we begin to explain how our powers of perception put people directly in the world. These two premises are implicit in the work of Dreyfus and Taylor which create a common endeavour to explore our humanity as a voice from nature. I then respond to those critics who focus on the notion of a voice of nature without fully exploring how Dreyfus and Taylor use language and perception to focus on what is environmentally significant for our lives. This argument is not at the expense of individual rights because the purpose is to understand our place in the world.

10
Taylor and Deep Ecology

For there is no science and no art of greater importance than
that which teaches seeing, which builds sensitivity and respect
for the natural environment, a world that 'has visibly been
recreated in the night'. A natural environment thus cherished
will always bring 'mornings when men are new-born, men who
have the seeds of life in them'.[1]

10.1 Introduction

This chapter draws together Taylor's evaluative and interpretivist framework and its connections with Deep Ecology. This involves a deeper analysis of the supposition that the natural environment possesses inherent
or intrinsic value. Taylor offers an understanding of Deep Ecology and
the political mechanisms needed to inform relevant publics without
drawing on problematic environmental metaphors. More fundamentally, Taylor argues that modernity is characterised by anthropocentric
attitudes, which assume that human beings are the final aim and end
of the universe.[2] Deep Ecology challenges this anthropocentric view of
humanity.

This chapter argues that the challenge motivating Taylor is to investigate the emancipatory and environmental potential in doctrines
such as liberalism, Deep Ecology and interpretivism. This investigation is evident in Taylor's work on Heidegger's philosophy in which he
explores our ability to perceive the natural environment and the values
that exist in it. He then connects these values back to Deep Ecology.
It is with interest that he has never fully developed these arguments,
thereby providing an opportunity to explore these connections. For
these purposes, the art of interpretation can be developed to explore

connections between environmental, political and spiritual values. Arguably, these values have been submerged in the anthropocentric quest to maximise economic returns in globalising world orders. The key idea to keep in mind is that Deep Ecology contains core values that democratic communities should consider, which Taylor takes further in his analysis of our procedural and secular age together with its impact on culture and the natural environment.[3] Taylor's environmental arguments examine not only how processes of disenchantment have been associated with the rise of modernity, but he also aims to create the preconditions to reform political structures.

For Taylor, the first task involves explaining the adverse impacts of modernity and economic development which reflect disenchanted conceptions of the natural environment. Modern processes of development continually search for escalating standards of living and growth but do not explore the price we pay for these high standards. From one point of view, Deep Ecology has created awareness that values exist in the natural environment which interpretivism relates to a valuing subject capable of appreciating and interpreting environmental values. From another angle, it is argued that Deep Ecology may create systems that protect nature over people.

Accordingly, the chapter is structured with three key sections. Immediately following this introduction is Section 10.2 that outlines some key connections between Taylor and Deep Ecology. Section 10.3 involves the provision of a succinct summary of Taylor's connections and differences with the claims made by Deep Ecology. Section 10.4 concludes the chapter by outlining Taylor's relevance for environmental politics.

10.2   Deep Ecology and environmental value

As mentioned in the introduction to this chapter, Taylor notes that he agrees with some of the axioms in Deep Ecology and also offers a steady-state and interpretivist political way of thinking. It is important to remember that key figures in the early Deep Ecology movement were philosophers such as Bill Devall, Arne Naess, and George Sessions, who emphasise the need for harmony and tranquillity in our interactions with the natural environment. Deep Ecological beliefs were in conflict with the prevailing social and political conditions of the time and struggled to gain political legitimacy. Early work in Deep Ecology, moreover, did not explain how its agendas would create peaceful political arrangements for achieving these goals. Taylor's framework is useful because his philosophical work has been committed to democratising aspects of Heidegger's critique of Western modernity that explored the role of place and belonging.

Taylor has stated that *Sources of the Self* is about the meaning and value in the natural environment, which was explored in Chapter 1 and 2 of this present book. This involves re-enchanting our world. Thus, Taylor concurs with Deep ecologists that we should search for value in the world, even though his approach involves work on language and perception taking our collective thinking in different directions. He observes that there are strands of Deep Ecology that usefully draw, for their very principles, from a variety of non-Western religions, ranging from Buddhism to Taoism to Native American Indian spirituality. These arguments in his work demonstrate that those strands of Deep Ecology are not necessarily opposed to the presence of religion in their own blend of home-grown spirituality as long as those traditions are not Western. Bron Taylor, Professor of Religion and Nature at the University of Florida, has observed:

> The existence of this perennial philosophy – a global religion of primal, ancient lineage, encompassing diverse, nature-benefi-cent cultures and lifeways, surviving especially among the world's remaining indigenous peoples, and still expressed in religions origi-nating in the Far East and in Jewish, Christian and Muslim mysti-cism – was an important assertion in the 1985 book Deep Ecology, edited by Bill Devall and George Sessions, that helped launch the movement. Although Naess, Devall and Sessions insisted that persons can arrive at the central convictions of deep ecology apart from any particular religious orientation, their own deep ecological ethics are clearly based on spiritual experiences in nature.[4]

Furthermore, the philosophy broadly decries the notion of people as self-functioning individuals with a sense of ethics and a moral self, instead positing mankind as a self-in-Self, a merely problematic actor in a greater chain of being that includes the natural environment. In such a vision, humanity operates from a sense of intuitive and spon-taneous situationally based morality rather than from a collection of pre-existing and rationally derived principles about how people should behave. Bron Taylor explains that Earth and nature-based spir-ituality is proliferating globally, which overlaps with Taylor's work on countercultural movements and their association with other religious movements.

Both Bron Taylor and Charles Taylor share the view that people can find ultimate meaning and transformative power in nature and that these experiences of nature-based spirituality are 'evoked by practices as diverse as mountaineering and neo-shamanic ritualising'.[5] These observations overlap with Taylor's *A Secular Age*, in which competing spiritualities can be integrated into a mosaic that transcends the limitations of our instrumental and secular practices. This reflects the central theme of this present book that Taylor's interpretivist work can be used to develop connections with political movements associated with Deep Ecology and other progressive movements. There is an implicit theism in Taylor's work that recognises the need to re-enchant our appreciation of the world in all its beauty and richness. Taylor states that he has found this in Deep Ecology and other environmental positions. Yet, he differs from those schools of thought by emphasising the promotion of better interpretations and direct engagement in the world which is discussed later in this chapter. For him, it is the philosophy of perception, which is expressed through our languages, which provide political avenues for improving attitudes towards the natural environment. Deep Ecology promotes a certain lifestyle and set of values.

Accordingly, Deep Ecology is critical of grand narratives, such as those espoused by liberalism and socialism, which assume that humanity can control and scientifically shape the world in its image of economic growth. Devall and Sessions argue that:

It's easier for deep ecologists than for others because we have certain fundamental values, a fundamental view of what's meaningful in life, what's worth maintaining, which makes it completely clear that we're opposed to further development for the sake of increased domination and an increased standard of living. The material standard of living should be drastically reduced and the quality of life, in the sense of basic satisfaction in the depths of one's heart or soul, should be maintained or increased... There is a basic intuition in deep ecology that we have no right to destroy other living beings without sufficient reason. Another norm is that, with maturity, human beings will experience joy when other life forms experience joy and sorrow when other life forms experience sorrow.[6]

Arne Naess, together with Devall and Sessions, outlined eight core principles of the Deep Ecology manifesto under which humanity was expected to live. The above argument illustrates that all life on Earth has intrinsic value, that humans have no right to reduce the richness and diversity of life except to satisfy vital needs and that the flourishing of nonhuman life requires a decrease in human population.[7] The principles demonstrate a world-view that recognises value in the natural environment. They romanticise pre-industrial and subsistence lifestyles while emphasising interconnected organic relationships between people and nature. The recommendations to reduce how humanity impinges on the natural environment are similar to Taylor's call to create steady-state political arrangements.

Yet, in re-enchanting political and philosophical world-views, Taylor aims to avoid some political pitfalls associated with the mystical elements in Deep Ecology. By placing nature's higher value at the centre of all deliberations, it has been argued that Deep ecologists have not fully engaged modern political structural problems. Deep ecologists have claimed to speak for the interests of nature, which may lead to traditional problems associated with those who claim moral authority to promote external reasons. A standard criticism of Deep Ecology involves the argument that it tends to impose its world-view on citizens and promotes illiberal political conceptions of positive freedom. Tom Regan, for example, has found this in Leopold's land ethic:

> The implications of [the land ethic's moral precept] include the clear prospect that the individual may be sacrificed for the greater biotic good, in the name of 'the integrity, stability, and beauty of the biotic community'. It is difficult to see how the notion of the rights of the individual could find a home within a view that, emotive connotations to one side, might be fairly dubbed 'environmental fascism'. To use Leopold's telling phrase, man is 'only a member of the biotic team', and as such has the same moral standing as any other 'member' of 'the team'. If, to take an extreme, fanciful but, it is hoped, not unfair example, the situation we face was either to kill a rare wild flower or a (plentiful) human being, and if the wild flower, as a 'team member', would contribute more to 'the integrity, stability, and beauty of the biotic community' than the human, then presumably we would not be doing wrong if we killed the human and saved the wild flower.[8]

Regan's concern shares traditional liberal concerns with positive freedom.[9] These problems have bedevilled the interpretivist tradition associated with the work of Michael Sandel and Taylor.[10] Nevertheless, both Deep Ecology and interpretivists share the view that intrinsic value exists in the natural environment; however, Taylor argues that to command moral authority, there must be an associated valuing subject.

The Taylorian response then involves fashioning how collective and common values are needed if people are to lead full lives. This response is not to simply rank natural interests above human interests in a zero-sum game. Rather, Taylor's interpretivism involves recognising the existence of these dilemmas in the first instance. The task of political philosophy is then to find a means to resolve these incommensurable dilemmas and at the same time to respect human freedoms and the rights of all for a decent standard of living.

Yet Regan's criticisms do not fully engage mainstream politics and existing communities. Thus, his arguments can be extended because environmental ethics must also engage the quest to escape the iron cage of modernity that has created instrumental approaches that ignore the commonalities alluded to in Deep ecological thinking. Thus, interpretivism is not an epistemology in the traditional sense but provides narratives that challenge the complex relationships implicit in the cosmos and as applied to environmental politics. While interpretivists may reject the more mystical elements of Deep Ecology, they concur in the manner in which they distance their arguments from dominant Western ideology, which separates humanity from a deeper understanding of the world. Taylor argues that:

> The Enlightenment gave rise to a new kind of indignant protest against the injustices of the world. Having demolished the older visions of cosmic order and exposed them as at best illusion, and perhaps even sham, it left all the differentiations of the old society, all its special burdens and disciplines, without justification. It is one thing to bear one's lot as a peasant if it is one's appointed place in the hierarchy of things as ordered by God and nature. But if the very idea of society as the embodiment of such a cosmic order is swept aside, if society is rather the common instrument of men who must live under the same political roof to pursue happiness, then the burdens and deprivations of this station are a savage imposition, against reason and justice, maintained only by knavery and lies.[11]

In overturning old social structures, the Enlightenment asserted that humanity was responsible for arriving at solutions that were authentic and just. Our relationships with the natural environment were defined in terms of utilitarian calculi and rational principles of order. The Enlightenment promoted polite society and rational order to create human control over the natural environment. These ideas are taken further in *A Secular Age*, which argues that our social structures have been built on disengaged and disenchanted visions of the natural environment.

In recent work, Taylor explains that humanity is now confronted with stark choices about how to live better lives. He argues that these choices do not necessarily have to stem from cosmic orders or Enlightenment principles of rationality. It is for this reason that he explores Deep Ecology and the way it aims to re-enchant our understanding of the natural environment. On this point, it is important to remember Taylor's work on practical reasoning recognises that environmental dilemmas involve understanding the connections with existing social systems and belief systems. Central to his framework is a discursive and political means to reconnect humanity with other values. This argument reflects the supposition that people are not independent, isolated entities who can provide a full portrait of their identity as it relates to the cosmos and the natural environment.

Therefore, Taylor discusses the need for negotiation and an appropriate use of language in promoting discursive approaches to the formation of knowledge and understanding. In these matters Taylor offers useful political ideas for the exploration of a deeper green community. He refers to the problems of modernity as a rupture between humanity and the values in the natural environment, which reflect the policies that we promote and the cultures we create. Interpretivism extends Deep Ecology and is critical in this regard as it involves evaluative frameworks that inform community deliberation through face-to-face conversations in dialogic societies. The aim is to set out the preconditions to understand our place in the natural environment. Taylor explores Deep Ecology through his interpretivist framework thereby engaging with more extreme forms of radical politics – the search is to alert us to the meaning in nature and deepen our lives.

10.3   Taylor's extensions to Deep Ecology

Taylor examines Deep Ecology through his work on language, perception and theism. Taylor uses Heidegger's work on language in *Being and Time* as a basis to explore the power of language. Indeed, Heidegger has inspired some Deep ecologists, whose work influenced and shaped much of the movement's early development. These concerns can be seen in the work of Devall and Sessions, as well as Naess, who develop a critique of Western philosophy that dates back to the time of Plato.

Enlightenment thinking espoused domination over nature, but from interpretivist quarters we are encouraged to think in terms of 'letting things be' and dwelling authentically in the natural environment with alertness and respect for natural processes.[12] This approach offers a means to explore why key environmental thinkers such as Michael E. Zimmerman have now abandoned Heidegger as an inspiration for their visions of the natural environment.[13] Zimmerman has abandoned Heidegger's philosophy because of its affiliations with National Socialism in 1933. Zimmerman has noted that the appeal of Heidegger's philosophy is its critique of capitalism and communism. It is through philosophy that environmentalists found new beginnings in Heidegger's philosophy and an opportunity to explore ways in which to transcend the limitations of economic rationalism and other instrumental approaches. For Zimmerman, Heidegger's limitation was his unwillingness to abandon his philosophy's 'inner truth' and 'greatness', which could infect environmental arguments. These environmental arguments have been criticised for imposing their values on others, thus returning to the limitations with positive freedom that I have discussed in earlier chapters.

Furthermore, Zimmerman distances himself from National Socialism even though his environmental arguments concur with aspects of Heidegger's work on technological modernity. Questions raised by Zimmerman's work lead us to ask questions such as: how do we live authentically in a technological age and what would our world look like if we lived in harmony with nature? This argument, of course, assumes two key issues. First, we are not living in harmony with nature, and second, that it would somehow be possible for us to achieve that harmony with nature. What would living in harmony with nature look like? How would we experience it? How would we know that what we perceive as being in harmony with nature was in fact accurate? Of course, Heidegger did not promote the worship of nature which might be found in some extreme forms of Deep Ecology. Heidegger defined nature in a way foreign to naturalism but did emphasise the German Volk as the source of new ideas and new beginnings.[14] It is, however, useful to re-interpret and revisit his philosophy because of many problems in environmental ethics.

Here Taylor usefully rethinks how Heidegger's work on language and realism can be used to restructure modern economic and political structures so that people might lead better lives in the natural environment. However, from Taylor's interpretivist quarters the arguments made by Deep ecologists, such as Naess and Zimmerman, offer environmental generalisations which seem to require political engagement with existing institutions. This is something Taylor does in his work on the public sphere. In that arena the environmental generalisation that can be discussed involves whether nature's value overrides other political considerations. This problem of engagement and generalisation is precisely the philosophical error that Deep ecologists find in Western political thought – they offer a critique of modernity without fully worked out philosophical strategies.

This thinking influences his affiliation with Deep Ecology to explore our place in the natural environment: as I have noted he uses a transcendental analysis of our being-in-the-world.[15] Further, Taylor's early work relies on transcendental arguments to offer an ontological examination of the meaning contained in the world. He argues that while intrinsic and objective values may not be real in an objective sense this does not mean that they are not receptacles of meaning. The argument is that if we adopt a post-Galilean stance we end up with a particular conception of the person – and an inherent limitation is to define rationality in such a way that it can only be conceived along scientific lines. We have lost sight of the meaning and splendour that many people find in nature. He continues:

> It is hard to see values as part of the ultimate furniture of the universe once one has adopted a post-Galilean stance and no longer understands the cosmos in terms of meaning. So we can be tempted to see our attribution of value to things as a kind of error, systematic and understandable, but error nonetheless. This was Mackie's position. Or we can understand these attributions as a kind of projection we make on the world...There is a sense in which values are not part of the 'objective' universe, but this cannot justify our considering them less real for all that. Values are not 'in the universe' in the sense that, if we were absent, values attributions could get no purchase on it. That is because an account of what are attributing, of various value properties, like 'attractive' and 'dangerous', involves at some level or other some reference to their impact on us.[16]

In the above passage, Taylor offers a way to think about the values we find to be of significance for us. His framework is not constrained by rational, technical and scientific thinking, and this is evident in his reviews on John McDowell's *Mind and World*. This argument is that we require a different way of thinking to reveal nature's value. Underlying his work on cosmic, cultural and environmental value is his metaphysics which involves taking seriously the issue of external values whose source is a power greater than ourselves, however that may be conceived.

From the theistic dimension in his work, Taylor uses these arguments to strengthen his Deep ecological views by exploring the value in nature in relation to a 'beyond' which involves two key dimensions. He emphasises the power within transcendence to experience the sense of fullness and wholeness that rare experiences of transcendence provide. Second this involves contact with a form of goodness beyond ordinary human life and whose source is a power greater than ourselves, however that may be conceived. Taylor's metaphysical approach connects with his work on the notion that objective value exists but not in the same sense that entities exist in the real world.

And the crucial point is that Taylor is attempting to explore how nature and meaning impact on our lives which involves the sense that there is some good higher than, beyond human flourishing. For Taylor this involves 'how we think about agape – the love which God has for us – and which we can partake of through his power'.[17] This involves the possibility of transformation which takes us beyond merely human perfection. As Daniel Baird succinctly points out 'this notion of a higher good as attainable by us only makes sense in the context of belief in a higher power, the transcendent God of faith which appears in most definitions of religion'.[18]

Therefore, Taylor offers an approach which concurs with Deep Ecology but also emphasises the role of theism and transcendence which must be explored in the various spheres of public life – Taylor mentions economic, political, cultural, and educational – are based on instrumental reason, the means by which they most effectively achieve their goals. As many critics have noted, Taylor's theism and 'his vast knowledge of the history of religion and theology is mostly rooted in Christianity', but he is careful to define religion in the broadest possible terms. He emphasises the movement towards secularism in the public sphere and traces how we have moved from a society in which it was virtually 'impossible not to believe in God, to one in which faith, even for the staunchest believer, is one human possibility among others'.[19]

At the very least, Taylor's Hegelian and Heideggerian arguments require rethinking how our sources of the self are also empowered whether through the cosmos, nature or a God of faith. The interpretivist perspective, moreover, maintains that it is important to consider these processes through which citizens are provided with opportunities to learn about the good and the value of the land through democratic discussion. I have stressed that the interpretivist perspective, however, emphasises that it is important to consider the processes through which citizens are provided with opportunities to learn about the good and the value of the land through democratic discussion. He contrasts an enchanted perspective with what we have considered to be one which is exclusively human. Whereas in exclusive humanism our ends and values are entirely oriented towards human flourishing – 'having a decent, comfortable life in reasonably just communities where people can fulfil their own potential, for instance – the religious point of view entails the possibility of higher ends and aspirations that transcend ordinary human life'.[20]

10.4 Conclusion

This chapter is built on explaining why Taylor concurs with the arguments of Deep ecologists that humanity should learn to live from the land but would allow the public to articulate its ecological concerns.[21] Taylor certainly avoids the tendency, common among Deep ecologists, to rank ecosystem interests higher than those of humanity based on a personalised conception of spirituality.[22] The complexity of Taylor's interpretivist philosophy involves grasping the connections between the particular in relation to the totality – where the particular relates to our connection with the totality of the natural environment.

In these various arguments, Taylor's interpretivism extends Deep Ecology to provide a deeper understanding of the way we impinge on the natural environment. He is aware of the need for sustainable solutions, which is evident in his early work on steady states. The implication is that human communities must be aware of ecological balance which invite us to at least consider issues associated with overpopulation and resource depletion that are becoming more problematic. Interpretivists, such as Taylor, emphasise that we must create new ways of living in the natural environment, but we cannot achieve this when we rely on philosophical approaches that rely solely on scientism together with its separation of our minds from the natural environment.

Although criticised for proposing a 'mysterious' view of nature, Taylor affirms that nature is not to be disenchanted. According to Taylor, enchanted nature points us to the ultimate *purpose* of creation where he follows Heidegger into a *Lichtung*, a linguistic clearing space where the mortal and the divine meet each other.

While it is outside the scope of my analysis to develop Taylor's theism it is important to remember that underpinning his work on Deep Ecology and the value in nature is a transcendental analysis of our being-in-the-world. Therefore, I have used Deep Ecology in this chapter as a point of comparison with Taylor's work on religion, theism and the transcendental dimension. In sum, I have attempted to explore how Taylor moves between Deep Ecology and his explicitly Catholic theological viewpoint. Taylor does not rely on the term 'theism' himself but it is important to remember that his view of 'theism' is used in the post-Enlightenment for a belief in a creator God. Taylor's transcendental analysis opens us to the view that there is something ineffable about the divine but that it is worth exploring how the world is saturated with meaning that is beyond the tools of empirical inquiry to understand. In this way Taylor moves beyond Deep Ecology by deploying a religious language that aims to open us to an approach which all members can relate to, and for this reason Taylor cleaves to a universalist approach.[23]

11
Taylor's Evaluative Framework and Critical Perspectives

This chapter examines how Taylor moves beyond critical approaches that are inspired by the work of Karl Marx. A number of critics have argued that Taylor's framework is unable to create environmental and social change. These criticisms overlook how he has moved from a humanist Marxist perspective to an interpretivist framework. His interpretivism questions the idea that social change can be created by revolutionary vanguard parties. Taylor's framework recognises the inescapable political difficulties in creating change and offers an approach which sets out some necessary preconditions for its creation.

11.1 Critical alternative futures

Taylor's political and environmental position has been subjected to searching critiques in recent years.[1] His critics claim that he relies too heavily on conversation and language to create environmental and social change. For example, critics such as Habermas argue that Taylor transfers the politics of species preservation to cultural and environmental politics and therefore lacks critical focus.[2] Whilst other critics such as Wendy Brown[3], Ian Frazer,[4] Jürgen Habermas argue that Taylor's interpretivist framework is not sufficiently critical of capitalism.[5] In this chapter I focus on Brown's criticisms to explore how Taylor deflects these concerns by exploring how interpretation can lead to the development of a new environmental awareness. This awareness emerges in the intersection between critical thinking, social imaginaries and interpretivist frameworks. These are speculative thoughts designed to recognise both critical and environmental sources of value which explore how nature is given consideration in modern social imaginaries.[6]

More particularly, a critical environmental ethic is implicit within Taylor's critique of modernity and is about developing the preconditions and conversations needed to create change. For Taylor this conversation is guided by a critical and ontic dialectic which emerges from his work in the Hegelian tradition. It is built on classical Greek sources and involves focusing on our being, culture and traditions. Therefore, Taylor's ontic dialectic offers a different way to consider the value of our social and natural environments. This is to be achieved through a political process that shuttles between our interactions in nature and our alternate visions for our communities.[7]

In exploring whether Taylor's interpretivism provides a critical focus, a return to his analysis of Marx may increase our understanding of our environmental and social worlds. But before we enter that arena it is important to keep in mind that Taylor goes beyond Marx to develop a politics of encompassment which brings citizens together. His key aim is guided by a vision of an environmentally aware society which recognises significant values for citizens and their communities. It is then argued that the analysis of capitalism bequeathed to us by Marx, and arguably developed by Taylor, explains how the 'self' and 'nature' are constructed not only through social labour but also the communicative and public spheres of civil society. Taylor's environmental conversation involves challenging a type of exclusive humanism that grips modern capitalist society.[8]

Taylor's interpretivism and critical approach might be our last line of defence against the globalisation of capitalism. This last line of defence is developed in five key sections. Section 11.2 introduces Taylor's work on dialectics in a critical context. Section 11.3 explores the issue of method within dialectical approaches as used by Taylor in his engagement with the work of Marx. Section 11.4 examines Taylor's work on humanist Marxism to offer a critical interpretation about how humanity adversely impinges on nature. Section 11.5 offers some directions for environmental politics and Section 11.6 concludes the chapter.

11.2   Interpretivism, dialectics and nature

Sir Isaiah Berlin has argued that Taylor seems to have been influenced by Marx's critique of capitalism, including the view that people can only rise to their full potential, if human society is liberated from the alienation, fragmentation and domination created by capitalism.[9] Berlin argued that both Taylor and Marx share a telos that spells out the vision of a good society where people act collectively and interact in free communication and discourse. Berlin's interpretation of Marx and Taylor is:

> that human beings can only rise to their full stature and develop all the potentialities which belong to them as human beings, if human society is liberated from oppression, exploitation and domination, which are inevitable consequences of, indeed, embodied in modern capitalism, but with their roots in various formations in the past.[10]

Responding to Berlin, Taylor commented on every point except the one concerning his relationships with the work of Marx.[11] This is a fascinating point because Taylor is highly critical of Marx in his classic book *Hegel*.[12] If Berlin is right to insist that Taylor is attracted by Marxian critiques of capitalism then there is an opportunity to develop this view in his interpretivist commitment to 'strong evaluations'; thus, providing an opportunity to investigate how interpretivism sets out the preconditions for reconciling societies with their natural environments.

Taylor's work on critical and environmental theory is based on his dissatisfaction with Marxian notions of the 'falling rate of profit', 'the collapse of capitalism' and 'the rise of the proletariat'.[13] According to Taylor, Marx did not fully consider identity, language or the sources of the self; shortcomings which require philosophical and also political intervention. Of course, Taylor acknowledges that revolutionary change is difficult to implement, but he emphasises that political change must be democratic. His criticisms of Marxism centre around his concern with its reliance on a proletarian vanguard framework which has the potential to impose values on citizens. Taylor's criticisms of Marx focus extensively on the latter's political writings where the proletariat is considered to be the agent in creating social change.[14]

Further, in Taylor's early work he suggests that Marxism's reliance on a communist blueprint and the proletariat leads to the will of the Party being imposed on citizens.[15] Taylor lays the blame on Marx's alleged utopianism where the communist model offers a blueprint that results in an unalienated society. He also argues that this problem of alienation also plagues other political systems where authoritarian rule is replaced by bureaucratic élites.[16] Taylor argues that political change must be created through democratic channels which aim to empower citizens through participatory structures. He does not offer all encompassing political solutions because his interpretivist framework is about creating links between particular values such that we begin to explore different features of the value in nature. This is to be achieved through a 'universal' commitment to equal facilitation between different cultures. This is why Chapter 8 focuses on Taylor's theory of language and communication. Furthermore, central to Taylor's interpretivism is the search for democratic public spheres that will inform citizens to determine 'what is to be done', not the Party. The role of the public sphere, therefore, is critical in the development of an environmental politics which is to be mediated through democratic interchange in communicative public spheres. The aim is to create common ground in civil society that heals divisions and respects others who are different from us. In this regard Taylor, is well aware that capitalism challenges, absorbs and deflects different political proposals as will the state, but maintains that change brought about by a vanguard party is illiberal.[17] Taylor, as a consequence, is critical of Marxian notions of communism which he believes will lead to undemocratic, vanguard politics.[18]

Rather, Taylor's interpretivism focuses on defining people as strong evaluators and this sets out a comparison with the objectives of meta-narratives such as liberalism and socialism. Earlier chapters showed that the simple weigher – the weak evaluator – focuses on appetite, whereas the strong evaluator examines the different possible modes of being of the agent. For the strong evaluator, it is not only desires that count, but also the kind of life and type of society that we want to live in. These are important discussions that must be given political consideration, such that ecological values are not kept at the periphery but reflect the kind of beings we are. Thus Taylor does not reduce human agency to a focus on modes of production or historical materialism.

As Taylor states, his aim is to take us to the centre of our existence as agents. He does this by exploring modernist meta-narratives such as liberalism and socialism[19] by explaining our strong evaluations and hyper-goods.[20] Taylor's work on strong evaluations and hyper-goods is designed to create accountability and freedom which he believes can be submerged in the socialist quest for an unalienated society. Further, earlier chapters have shown that hyper-goods are those shared goods in the community which reflect the values that are worth pursuing in a democratic society and which could, for example, be committed to pursuing reconciliation between the natural environment and our civil societies. For environmental politics, hyper-goods are not only incomparably more important than other goods, but also provide the background context for making evaluative choices.[21] Using Habermas's terminology these are the goods to which we turn in times of crisis. Taylor acknowledges the existence of hyper-goods to emphasise the 'moral' in our culture and focus on the issues most important for our existence. Taylor's work, through the notion of hyper-goods, therefore offers a critical stance towards modernity and also offers a means to think politically about the political structures needed for environmental politics.

11.3   The issue of method and framework

Taylor reconsiders the role of critical philosophy which is guided by his quest to explore our place in the cosmos and in nature. As I argue in Chapter 9, Taylor's early work explores these issues by studying various approaches to dialectic methods to examine how we come in contact with nature. This is undertaken by exploring three dominant dialectical approaches which this chapter deepens.

- The first school of thought claims that the modern scientific tradition refutes dialectical methods as mystical and subjective.
- The second school of thought explores a political tradition which relies on a dialectical representation of reality, culture and nature.
- The third school of thought uses an ontic dialectic to investigate culture, history and politics according to notions of 'strong evaluation'.[22]

According to Taylor, Western political philosophy has taken the first direction and rejects dialectical methods as unscientific.[23] The second school of thought encompasses orthodox Marxism and sees the structure of reality itself as fundamentally dialectical. The first two schools of thought are diametrically opposed.[24] The third school broadens the dialectic to understand human beings, their lives and history. Taylor argues that the third school does not debunk the scientific tradition but it does maintain that dialectics should not be established as a category of reality as such, in the way it is established, as a central dogma of dialectical materialism.[25]

Taylor classifies Marx's work on capital, the falling rates of profit and historical materialism as falling within the scientific tradition and so inevitably narrows the role of dialectical logic. He argues that Marx's work on capitalism presents a 'disaster scenario' and that this is because the capitalist social system focuses excessively on economic growth.[26] This reflects Marx's mechanical methodology which leads to a focus on the falling rate of profit and an anticipation of the eventual demise of capitalism. Taylor claims that Marx ends up relying on debatable scientific methods to support his conclusions about capitalism. Marx and Marxists, argues Taylor, rely mechanically on testing whether the 'rate of profit declines', or 'capital agglomerates in ever larger masses', and whether 'communism is the historical culmination of human society'. Taylor concludes:

> This transcending of external determinants in favour of free expression makes perfect sense, of course, in the categories of Hegelian philosophy. It is what we see when we ascend the levels of being in Hegel's philosophy of nature, e.g., from inorganic nature to life, or when we move from alienation back to *Sittlichkeit*. But it is foreign to the established tradition of science rooted in the Enlightenment. This can allow for man objectifying nature as a domain of neutral instruments on one level while he shapes an expressive boundary whereby what is in the realm of objectification and natural law at one stage of history is pulled beyond it into the realm of expression at another. It makes no allowance for its laws being *'aufgehoben'*.[27]

In this passage, Taylor illustrates the limitations of a type of scientific approach which treats the natural environment as an object of neutrality. Capitalist societies are not able to create a dialectical *'aufgehoben'*, where *'aufgehoben'* refers to the processes of transition from one stage to another. The Marx-Engels tradition, according to Taylor, ignored the internal avenues of communication in communities and overemphasised the economic and material imperatives in capitalist crises.[28] For present purposes, Taylor's argument implies that Marx goes some way towards explaining these issues but the tendency to search for law-like regularities (wrongly) assumes that these discoveries can be unproblematically applied to human affairs and environmental issues. In other words, the conversation about our place in nature is not legitimately foreclosed by invoking the law-like regularities of capitalism and so remains open-ended.

Consideration of Taylor's interpretation of Marx's dialectical perspective impresses just how important the role of frameworks is in guiding how we implement strong evaluations and their place in society. When thinking about the natural environment it is striking, in this regard, how persistent social and ecological concerns circle around work on the ontic dialectic. According to Taylor, the method of Marx is overly reliant on the notion of commodification and processes of historical materialism.[29] This line of thought is challenged by Wendy Brown in her claim that it is Taylor who glides over the contradictions in capitalism without recognising their basis in historical and material processes.[30] She argues that in overlooking the power of Marx's analysis it is interpretivism which fails to challenge how capitalism creates an exploitative and technical stance towards the natural environment.[31]

However Taylor develops the spirit of Marxism through an ongoing conversation of humanity that has no ultimate end because it presents a challenge to the politics of socialism – which can submerge the meaning and intrinsic value in the natural environment in its focus on social labour. Marxists try to prove that capitalism follows law-like regularities but Taylor believes that this tends to obfuscate the pluralism that exists in modern communities. In other words, for Taylor Marx's version of socialism replaces capitalism with a system that tends to foreclose debate about how modern cultural, environmental and religion place new demands on our thinking about the world.

11.4   Taylor's critics and the role of Humanist Marxism

This section of the chapter explains why modern Marxists such as Wendy Brown and Ian Frazer believe that Taylor is too quick in dismissing the relevance of Marx. In particular, Brown believes that Taylor overlooks how material forces shape our historical configurations. That is, historical and material forces shape the social spheres in which we live. Moreover, Brown emphasises how materialism guides Marx's critique of capitalism and its association with religion. Brown continues that Marx's approach is superior to that of Taylor's analysis of our secular age because it is capitalism that absorbs and deflects attempts to reform it. Moreover, Brown claims that Taylor devotes limited space to the work of Marx in *A Secular Age*; however, previous sections of this chapter have already illustrated Taylor's extensive engagement with the Hegelian and Marxian dialectical method. Of course anyone who has ever attempted to come to terms with Marx and Taylor will understand how impossible it is to cover major aspects of their thought in just a few pages. In this chapter the aim is simply to introduce provisionally some Taylorian motifs to respond to some critics whilst exploring our place in the natural environment.[32]

Methodology: *The Grundrisse*

Taylor's interpretivism moves beyond modern Marxists who often refer to Marx's introduction to his *The Grundrisse*. The *Grundrisse* comprises the reading notes that informed the monumental *Capital*. In these reading notes Marx offers an approach which engages with Hegel's methodology. For Hegel the ontological starting point was mediation. The notion of mediation posited a connection between matter and spirit (*Geist*). In *The*

Grundrisse Marx seemingly inverts Hegel's view such that what appears to be immediate consciousness turns out to be mediate. Marxists then claim that this approach creates an appreciation of our existence only as part of a whole defined in the way humans interact with nature and with each other – interactions defined as modes of production – in producing the means of life.[33] This is according to the processes of historical materialism where human society changes through the means by which humans collectively produce the necessities of life.

Many theorists find the clearest example of Marx's method of inquiry can be found in *The Grundrisse*. Marx points out that there are no ontologically prior categories but suggests that each category must be related to the whole of which it is a part. Marx illustrates this in his famous analysis of population as the foundation and the subject of the entire social act of production:

The population is an abstraction if I leave out, for example, the classes on which it is composed'. These classes in turn are an empty phrase, if I am unfamiliar with the elements on which they rest. Eg. wage, labour, capital, etc. These latter in turn presuppose exchange, division of labour, prices etc. For example, capital is nothing without wage labour, without value, money, price, etc. Thus, if I were to begin with the population this would be chaotic conception [Vorstellung] of the whole, and I would then, by means of further determination, move analytically towards ever more simple concepts [Begriff] from the imagined concrete towards ever thinner abstractions until I arrived at the simplest determinations. From there the journey would have to be retraced until I finally arrived at the population again, but this time not as the chaotic conception of the whole, but as a rich totality of many determinations and relations.[34]

For Marx, understanding a social system involves breaking a social whole into simpler elements and then relating them to each other. Here, Marx is consistent with Hegel in breaking down the various elements of society and reassembling them to show the 'rich totality of many determinations and relations'.[35] Marx's method explains how a category of simple unity proceeds to more complex formations and back again thereby locating the material forces that shape human communities. It resembles Hegel's dialectical method where 'the beginning has for the method no other determinateness than that of being simple and universal; this is itself the *determinateness* by reason of which it is deficient'.[36]

My argument is that whilst Marx's method may resemble that of Hegel's dialectic, it is Taylor who draws our attention to the dialectical implications of our cosmic, environmental and social forces. These factors are critical in shaping our communities. Thus, Taylor is not denying the importance of the forces of historical materialism, as suggested by his critics, but is examining our place in the cosmos and in nature. Further Taylor's ontic approach to social theory is endeavouring to understand a society in which people are free to criticise its institutions and try to overcome its contradictions through ongoing debate. This is part of his ongoing conversation which has no ultimate end and is about ever widening its openness and structure. That our way of thinking is dialectical is essential for its deeply critical constitution; that it accepts its interpretive limitations is essential for a truly open debate.

Marx and nature: the commodity form

It is often said by environmentalists that Marx had little to say about the natural environment. Modern Marxists such as Brown and Frazer respond by locating many passages in the work of Marx which illustrate a sensitivity towards the natural environment. Moreover, Marx uses this method to conclude that under capitalism the wealth of communities is the sum total of commodities, where wealth is related to utility as use value. It is therefore important to remember that Marx explains:

> The total labour-power of society, which is manifested in the values of the commodities, counts here as one homogeneous mass of human labour-power, although composed of innumerable individual units of labour-power. Each of these units is the same as any other, to the extent that it has the character of a socially average unit of labour-power and acts as such, i.e. only needs, in order to produce a commodity, the labour time which is necessary on an average, or other words is socially necessary.[37]

Under capitalism consumers and producers are brought together only through the exchange of commodities, whereby the intrinsic value in nature is submerged in the search for profit. Marxists however continue to focus on the commodity form as the key to understanding capitalism – the commodity is assumed to be logically prior to the form of *abstract labour* (which is the labour common to all commodities and creates a system which reduces value to labour content).[38] In capturing the essence of capitalism, then, we are provided with an opportunity to explore what is nature and what nature means. Of course, in other parts of his work, such as his *1844 Manuscripts*, he explains the importance of the laws of nature in understanding beauty and value. He states:

> An animal forms objects only in accordance with the standard and the need of the species to which it belongs, whilst man knows how to produce in accordance with the standard of every species, and knows how to apply everywhere the *inherent standard* to the object. Man therefore also forms objects in accordance with the laws of beauty.[39]

In passages, such as these, Marx shows his sensitivity towards beauty and nature where standards of wealth *are actually in the minds of the beholder and are not simply created from out of thin air.* Marx argues:

> Every child knows that any nation which stopped work – I will not say for one year – but just for a couple of weeks, would die. And every child knows that the volume of products corresponding to the various needs calls for various and quantitatively determined amounts of total social labour. It is self-evident that this is *necessity* of the *division* of social labour production, but can only alter its *mode of appearance.* Natural laws can never be negated. Only the *form* in which those laws are applied can be altered in historically different situations. And the form in which this proportional division of labour asserts itself as a *private exchange* of the individual products of labour, is precisely the exchange of those products.[40]

Marx argues that commodities are transformed from natural endowments and are the source of value. Moreover, *In The Critique of the Gotha Programme*, Marx offers a perspective where nature is seen to be the source of wealth and implicitly of value.[41] From a Taylorian perspective it seems that nature is reduced to a commodity – and the commodity form replaces earlier forms of social cohesion found in religious practice. Of course, as Brown argues in passages such as these Marx's analysis is not reductionist. But Taylor's argument is about how religious consciousness opens us to the cosmos or nature.[42] Thus, it seems that this investment in Marx is at the expense of an appreciation about how our interactions in nature empower our life-worlds.

Taylor's work on religion, perception and understanding is about creating space to explore the transcendent in such a way we open ourselves to the meaning and value in nature. Thus, Taylor's point is not to deny the importance of the commodity form or relegate it to the category of a religious symbol but to recognise that commodification cannot completely account for the significant relationships needed to empower us against the closed world forces of modernity. Thus Taylor emphasises how social cohesion and integration are achieved through the mechanisms of belief, language and transcendence in the differentiated form they take in modern society, which relies on the special communicative achievement of reaching understanding and agreement.

11.4 Marx and Taylor: interpretivism and environment awareness

In the preceding sections of this chapter, I have attempted to clarify how Taylor extends Marx's work. The aim is to create an environmentally and culturally sensitive society. Taylor is relevant to this task in the way he understands historical progress which Marx explained as involving progress from domination to freedom where the domination of the environment is a precondition to human history. Taylor warns that Marx's arguments, for example, in *The Communist Manifesto* could result in a technical and arid conception of community and the natural environment. Taylor refers us to Marx's argument that:

Communism as a fully developed naturalism is humanism and as a fully developed humanism is naturalism. It is the *definitive* resolution of the antagonism between man and nature, and between man and man. It is the true solution of the conflict between existence and essence, between objectification and self-affirmation, between freedom and necessity, between individual and species. It is the solution of the riddle of history and knows itself to be this solution.[43]

Here the Taylorian interpretation is that Marx maintains that the environment is to be technically controlled under communism. The implication of this argument is that Marx reduces nature to a commodity within a mode of projection. Of course, Marx does provide a technical view of nature; Reiner Grundmann reminds us of the example of 'Bavaria's sluggish peasant economy, the ground on which grow priests and Daumers alike, should at last be ploughed up by modern cultivation and modern machines'.[44] In passages such as these Marx's evident ambivalence towards the natural environment is evident – it seems Marx invites us to take seriously the historical and social conditions which have led to these perspectives. If, however, we scan the corpus of his work it becomes clear that Marx develops an awareness of nature's significance in human history but my argument is that these dilemmas are now more complex than ever before.[45]

However, whilst one might concur that Marxian inspired critics of Taylor assist us to understand the problems of capitalism they do not fully come to grips with his search to set out the preconditions needed to create social change.[46] This is a necessary task given the problematic implications of the reliance on the proletariat to initiate change (which may not be environmentally sensitive). Whilst I have demonstrated that Marx's awareness of the environment is not in question: it seems that the various viewpoints can be used to support various arguments. I argue that Marx points us towards not only a critique of capital based on historical materialism but opens us to a realist analysis of our place in the natural environment. I propose this commonalities strategy, in the spirit of Taylor's search for a politics of encompassment, to emphasise a communicative framework whilst also focusing on the notion that the natural environment is a strong evaluation. The ideas behind our strong evaluations shape people's being as they interact in the world and in this way Taylor moves beyond Marx to explore our full dimensions of being human. This is why earlier chapters explored Merleau-Ponty's work. Taylor reminds us:

> [Merleau-Ponty] departs from Hegel and Marx in that he rejects the necessity of the happy ending. History may never resolve the contradictions now at work in it. The development of history is profoundly contingent. The nineteenth-century belief in inevitable progress is a kind of 'retrospective illusion'. Once a new institution, idea, work of art, philosophy is born, it makes us see all past history as leading up to it.[47]

For the purposes of a critical ecological politics it is Taylor's recognition of contingency and the search to combine divergent goods in a finite polis which offers a unique approach to our understanding of nature. Moreover, Taylor's framework is guided by an enthusiastic commitment to political reasonableness and recognition – it guides our thinking that in our interactions with other communities we may have to give and support their interactions with nature. Put simply, it may not be totally possible to model these interactions using economic and materialist frameworks that glide over our more basic primordiality of our object engagement. As I have argued Taylor does not maintain that there is a totally coherent and universal solution to our political dilemmas. In recent work he explains:

I think we're still crippled by many of our policies. We are crippled with the idea that there is such a thing as socialism, a totally coherent solution to all political problems, and that if you think differently you're on the other side; either you're going further into socialism or you're betraying it. I think the way to conceive our situation is that we are living through a set of extremely painful dilemmas, and are going to go on doing so. In other words, capitalism is, if you like, the source of a number of terrible dilemmas. We can't live with it; we can't live without it. We can't live without it because it's tied to certain kinds of growth and so on.[48]

Taylor's critics have seemingly not explored these interpretivist dimensions in his work and do not fully recognise that this ontological quest involves coming to grips with a totally different way to think about our place in the natural environment. As I have argued this involves his work on direct engaged coping to explore our place in the world which leads us to a different way to define our interactions with others.

11.5   Towards a post-interpretivist and critical ecological politics

Critical ecological politics must consider nature's meaning and value as well as the social structures which mediate these interactions. This strategy, therefore, involves using Taylor's commitment to the development of commonalities broad enough to accommodate ecological and social values. These commonalities should be broad enough to accommodate insights from critical thinkers such as Wendy Brown, Ian Frazer, and Jürgen Habermas who have engaged the work of Marx. These writers continue to adapt Marx's insights to critically analyse the object under consideration; in this case capitalism and its impact on the environment. Taylor's work differs, however, from Marxism in that he emphasises communication and dialogue through ever expanding public spheres thereby providing space for counter-movements, other voices and perspectives to participate in community deliberation.

Furthermore, it is through critical public spheres that environmental awareness and change emerge in a new space which accommodates labour participation and the construction of improved relationships with the natural environment.[49] This is where the creative aspect of labour and labour process, in particular, have been submerged by the fundamental logic of capitalism. It is this instrumental logic, or malaise of modernity to borrow Taylor's terminology, which suppresses an awareness of nature's meaning and value. The development of a critical ecological politics can be explored through reference to the recent work of Timothy W. Luke. His environmental ethic has departed from the ideas presented by Marx. In particular, Luke has questioned the relevance of Marx for environmental politics and has argued that we must continue to search for new strategies to deconstruct modernity's political structures. It is modernity's structures, which are destructive of the natural environment. He argues:

> Under transnational corporate capitalism, the public goods of nature are increasingly despoiled for a historically limited group of individuals who are 'fortunate' enough to be alive as relatively affluent consumers in the twentieth century, at the expense of all other human, animal, and plant communities.[50] This consumerist social model, as packaged by capitalist megatechnics, thrives on the simultaneous denial and destruction of the ecological limits embedded in human habitats. The project of social ecology and voluntary simplicity arises from the practical challenge of elaborating a new model for ecologically sound communities that would coexist in diverse habitats without qualitatively lowering many modern living standards.[51]

Echoing Marx's critique of capitalism, Luke however, advocates a postmodern direction to create an escape path from modernity's anthropocentricism. Luke also emphasises a political need to reform humanity's anthropocentric stance to the natural environment. But, postmodern methods seemingly offer only deconstruction whilst downplaying the prospect of reconciliation which is based on their supposition that liberalism and interpretivism both impose one set of universal and foundational values on others.[52] A critical ecological politics, therefore, involves modifying and exploring relationships between Marx and Taylor. It is Taylor who reminds us that Hegel's dialectic builds a rich and multilayered picture that draws together through a unity-in-difference as a means to consider different ways humanity can be related to the natural environment.

A critical environmental ethic, taking a post-interpretivist pathway, explores the ability of capitalism to reproduce itself through a capitalist system that commodifies social labour. In utilising Taylor's developments in phenomenology a critical ecological politics aims to reconcile political differences in communities. Taylor, it will be recalled, criticises Marx's social and environmental perspective based on his interpretation that communism leads to a situation where 'generic man dominates nature and can impress his free design on it'.[53]

11.6 Critical ecological politics: concluding thoughts

This chapter has explored Taylor's work on a dialectic way of thinking and why he thinks it ought to be interpretivist in considering the factors needed for an environmentally aware and free society. That is where one understands it to be a society in which people are free to criticise its institutions and try to overcome its contradictions through ongoing debate, a debate which should take place among ever widening circles of more and more participants with equal rights. That our way of thinking is dialectic is essential for its deeply critical constitution; that it accepts its interpretive limitations is essential for a truly open debate. The false claims of dialectic science are damaging to such a debate since they cut off the dialogue of interpretations. That kind of science rather forms the basis for a totally different form of social practice whose development is intended to be controlled by a blueprint and invariably by an elite.

The scientific approach to dialectics goes naturally together with a technological model of society and with social control. I have used Taylor's interpretative dialectic to be the foundation for a dialogic model. Following Taylor's use of the language of Aristotle, it has been argued that an interpretivist concept of dialectics – an ontic dialectic – can form the basis for political regimes appropriate for their time whilst dialectic science is more likely to provide the foundation for a despotic regime. Taylor's development through Hegel and Marx therefore operates through consideration of different notions of human flourishing and modes of governance where adjudication is determined through principles of self-respect and dialogic reason. Taylor's notion of strong evaluations can be related to his work on the public sphere in order to determine the extent to which capitalism deflects critique and modifies reform proposals. He, therefore, advocates discursive structures to adjudicate between strong evaluations with democratic overlays.

12
Conclusion

12.1 Conclusion: the search for environmental commonalities

A central aim of Taylor's evaluative framework is for citizens to regain control over their lives in a political society where the public sphere provides access for citizens to participate in community values. I have argued that Taylor's interpretivism explores the need for a critical perspective on capitalism. Moreover, Taylor offers a discursive and interpretivist framework to explore how we might begin to tame the adverse forces of global capitalism. On this point, he argues that the public sphere must be refashioned to accommodate democratic interchange so that citizens have opportunities to reflect on their way of being-in-the-world.

Central to Taylor's work is a desire to expand the public sphere and for this reason recent work examines environmental politics, religion and processes of secularisation. In the early chapters I explore his work on the role of religion by asking 'what is secularity'? In an early piece he observed that when people talk about 'secularisation', they can mean a host of different things. In one sense, the word designates the decline of religious belief and practice in the modern world, the declining numbers who enter church, or who declare themselves believers. In another, it can mean the retreat of religion from the public space, the steady transformation of our institutions toward religious and ideological neutrality in their shedding of a religious identity.[1]

For Taylor, neither of these interpretations of modernity and its secular processes are attractive. He argues that there are various senses involved in understanding the connections between liberalism and secularism. These connections require analysis and interpretation if we are to improve our understanding of the natural environment. It is important to understand these senses if we are to not only understand how we have arrived at our present predicament, but how we might better manage competing world-views as they impact on the natural environment. This is because modern secular society has not managed the range of religious and spiritual beliefs that empower many people's vision of the world. Taylor argues that our ideological and political structures have become closed world structures which deny the existence of intrinsic value and meaning in nature. The economic focus on sustainability tends to close off ideas about our place in the natural environment and what our relationship to it should be.

12.2   Taylor, Merleau-Ponty and nature

Here I am inspired by the recent argument made by the political interpreter, Timothy Moyle, who outlines some of the environmental and practical implications associated with our direct engagement and the values revealed through our language. Moyle points out that while animals are expressive creatures and thus to an extent 'inspired', nonetheless only in humanity does inspiration entail giving thanks to life for the 'gift of life and thus realising the gift by *returning* it *to* life. It is only humans who can return the gift'. This distinctive mark of the human being is the highest form of animal expression: freely and actively to re-shape and modulate the (second) nature that gives to us the capacity to think. Moyle continues:

> In a Heideggerian variation on what is essentially Aristotelian thought, the human activity which most of all fulfils the reception and return of the gift of Being is thinking, the free exercise of reason embodied in the creation of a philosophical work. As Merleau-Ponty remarks, art and philosophy are not fabrications in the universe of culture but contact with Being (or presumably nature) precisely *as* creations: 'Being is *what requires creation of us* to experience it' (ViIn, 197).[2]

Moyle's interpretation of Merleau-Ponty's thinking points to new political pathways which involve a deeper ecological politics, one that offers empathy as opposed to transformation of the natural environment. Moyle uses the work of Merleau-Ponty and Heidegger to argue that people are directly engaged with the natural environment and not separate from it. Further, there is a certain continuity between Kant, Heidegger and Merleau-Ponty which begins by asking how the phenomenon of experience is possible in the first instance. In utilising insights from Merleau-Ponty, and also Gadamer and McDowell, the aim is to explain how our perceptual taking-in of already conceptually structured content puts us in touch with the natural environment.

Accordingly, humans are not thinking minds alone, but are engaged agents whose perception and senses put them in touch with nature. This generates a sense of awe in the face of nature, reflecting the fact that our interpretations and stories report only some of the meaning that the natural environment contains. The book argued that to challenge modern accounts of our place in the world involves examining why we emphasise one side of our experience, that of our human subjectivity. Here it is stressed that Merleau-Ponty explains the role our perceptions play in absorbing and reflecting the meaning and values in the world. Perception involves our total engagement with the external and natural world.

12.3 The issue of evaluation and method

I have also argued that when one compares the Marxian methodology with that of Taylor it soon becomes clear that Taylor's ontic dialectic proceeds in a similar but deeper way to develop a rich picture of the totality which includes the natural environment. In a manner similar to that of Marx Taylor, however, does not suggest that universal laws operate across time and space but maintains that we must be attentive to a particular phenomenon. In sharing a method that moves between observation and theory, it becomes possible to uncover changes over time in certain tendencies manifest in a mode of production. This way of thinking provides a means to think about the natural environment and reconcile our understanding of the social and natural worlds.

More particularly, Taylor's reading of Hegel and Marx thereby provides a means to avoid the criticism that Marx levelled at Hegel as too abstract. In this manner Taylor takes Hegel's work on reconciliation further in linking the particular with an element of the universal through dialogic social structures. I have used this strategy to consider humanity's impact on the environment, and the social causes of any such crisis. Environmental and political reform involve balancing different value systems such that the natural environment is recognised in all decisions. Here, Taylor's method shuttles back and forth between the particular and the totality. Taylor enables us to think about how we might build communities that work toward making people's lives free from aliena-tion, and respects the environment. It also points toward a possible solu-tion to the dilemma between humanity and the environment which

Taylor has said is at the centre of his attempt to rethink ways of human flourishing to transcend a procedural social system to another.

I have argued that Taylor extends and develops critical theory through his interpretivism thereby challenging the social imaginaries on which capitalism is built. I have argued that Taylor's interpretation of critical and postmodern approaches emerges from that tradition which criticises Marx as too abstract. It is for this reason that Taylor's broader epistemological and political stance toward the environment is used to reassemble the particular in the light of the totality. For ecological politics, the key point is to understand that the environment is a component in the totality through which humanity realises their being-in-the-world. Therefore, in revisiting political method I have argued that it is possible to clarify Taylor's commitment to 'strong evaluations' which has either been ignored by his critics, or possibly reinterpreted as a common good. The notion of a strong evaluation guides a social totality that therefore, relates individual categories to that totality while keeping sight of society as a whole, as well as relying on democratic methods to resolve different demands.

He provides a means to clarify how we live and relate with the environment. It becomes clear, therefore, that Taylor's notions of strong evaluations and hyper-goods enable him to contemplate a radical transformation of capitalism through republican political structures. In a republican polity, with dialogic overlays, it might be possible to constrain individual self-seeking, where the company form is assumed to be the means to effect social and environmental change in liberal and capitalist societies.

12.4   Interpretation and strong evaluations

Interpretivism, engaged coping and strong evaluations, must be vigilant in the way it theorises how capitalism moderates and absorbs different demands. Taylor's discourse theory is about shuttling back and forth between observation and theory to understand relationships. This way of thinking, therefore, does not advocate that definitive conclusions must be determined in the discourse arena. This is because it is citizens who decide what is desirable in the democratic polity, and is a consequence of democratic deliberation. The point of Taylor's ethic, therefore, is to derive a universal understanding of phenomena from a particular social totality where people are provided with opportunities to participate in the democratic process. Interpretivists respond by explaining how significant elements may form part of a potentially meaningful totality which contains *lacunae*, or missing pieces of evidence to be reconstructed through historical inquiry. Those significant elements may be fluid, but they are not purely subjective, and so cannot be approached in a relativistic manner. They may be ascertained in hermeneutic fashion – in the same manner as text. The Taylorian framework involves processes of practical argumentation in public spheres. Taylor assumes that 'my opponent already shares at least some of the fundamental dispositions toward good and right which guide me'.[3] 'The error comes from confusion, unclarity, or an unwillingness to face some of what he cannot lucidly repudiate; and reasoning aims to show up this error.'[4]

Through practical reasoning, unity and diversity are balanced in hermeneutic fashion whereby this approach acts to respond to the limitations of liberal neutrality and never-ending postmodern deconstruction. A further interpretive point is that realism and skilful coping are the direct means that put us in touch with the natural environment. This is based on the supposition that once it is realised that interpretation and significance are the key concepts which can be used to reveal and express intrinsic value in context. I have argued that Taylor makes an interesting hermeneutic move to explain how faculties of understanding utilise spontaneous capacities to interpret the world. I point out that through language the goal is to provide a best account which involves its highest fulfilment. Furthermore, the values we experience using our coping skill are shaped in the affordances of attack and defence (as reflected in debate and dialogue). Accordingly, our interpretive and scientific methods create richer interpretations through progressive enquiry: the issue is not just about nature's intrinsic value, but also our authentic interactions with it. The ideal of authenticity reflects our embeddedness with others as well as the natural environment. From this perspective, Taylor takes the interpretive debate further by maintaining that our practices can be aligned with authentic precision to reveal the value of nature – Taylor's ethic of authenticity is used to reveal the inadequacies of anthropocentric stances and how values are revealed through our practices.

It has been argued that instrumental reason leads to an economic world-view associated with individualism and negative freedom; that is, freedom to pursue whatever ends individuals see fit. However, there is more to freedom than radical autonomy and individualism might lead us to believe. Freedom also involves our relationships in communities and with the natural environment which we share with other people. It was for this reason that I argue that Taylor's environmental arguments overlap the axioms of Deep Ecology and concur that humanity's controlling and dominating attitude submerges other values that can be found in nature. Through practical reason people engage with the world and consider how perception plays a means in visualising the world. For the purposes of our awareness of nature such thinking provides a different way to consider its meaning and value. It is for this reason that interpretivists point to our second nature as it incorporates feeling, empathy and our perceptions of the world. It will be recalled that Dreyfus summarised this aspect of the interpretive framework such that we can only understand how we are normally in touch with the cosmos when we see that we are not disembodied, detached, contemplators, but rather, embodied, involved coping agents.[5] In recognising the existence of these perceptual and coping skills, it is then possible to respond to closed world thinkers to whom community and nature are likely to lead to external standards which limit freedom. Often critics overlook Taylor's realism and (anti)-meditational epistemology that overcomes the rigidity in our closed world political structures. It is a categorical error, therefore, to argue that Taylor offers an ethic of authenticity which relies on nature as an external standard; rather, his framework is about examining how different practices reveal different thinking and other ways to live in the world.

Interpretivism avoids imposing preconceived value on objects, and support for this argument can be found when he emphasises that the ontological level shapes, but does not determine the social, political and cultural goods that communities may advocate. Taylor visualises the world, rethinks our beliefs, and how our choices impact on the natural environment. He explains that what allows us to make the difference is a rich, largely implicit, and of course inherently contestable understanding of what the important meanings and purposes of human life are. As I have argued earlier often critics overlook the role of perception in Taylor's work. They do not explore how perception relates to common goods, language and background values as a part of a broader whole.

Of course, the various parts of the social whole are always open to reinterpretation and reconsideration. The natural environment, however, is a common endeavour that includes humanity and can be conceived as a text in which we live. This observation emanates from Hegel's argument that the mind is in the world and is not separate from that world. This proposition requires different political structures and ideas to meaningfully engage with humanity's being-in-the-world. Taylor develops these ideas expressing a political perspective that re-theorises politics, reconsiders humanity's place in the world, and restructures our sense of self in the world.

On Taylor's view, there is no reason why humanity has to re-shape the natural environment for economic purposes only. The interpretivist dimensions associated with human reason offer scope to design new and different means to relate with the natural world. There are no indubitable reasons why our communities should be limited by a path dictated by instrumental reason and unrestrained consumerism. Interpretivists begin their exegesis by explaining how our current practices have ignored how we are related to and shaped by the world we live in. They emphasise feeling and perception as having a valid role in the experience of embodied agents making their way in the world. From this observation, interpretivists argue that a scientific approach treats the natural environmental as a simple commodity and submerges other ways to value the natural environment. The application of a technical method to the humanities can submerge an appreciation of the moral frameworks more applicable to human sciences. Humanity's power of reason and rationality provide the human race with objectifying powers that transform the natural environment.

In sum, I have argued that interpretation is linked to a moral realism that assists our thinking about the politics of the 'environmental other' and 'the end of history'. Taylor's work can be used as a different pathway where commonalities are nurtured and sustained in a dialogic society. In such a view, authenticity is about being true to a role and understanding the world we live in. It is for this reason that Taylor's (anti-) epistemology defines agents as purposive beings who pursue their goals in a context imbued meaning. This is where engagement with the world involves coming to terms with how context influences the values we express. The interpretive strategy observes that the construction of knowledge is not just theoretical, *inter alia*, but involves interactions with others and the world. It is at this point that Taylor argues that our interpretations are a vital ingredient and medium through which knowledge and our understanding of the natural environment is developed.

Notes

1 Introduction

1. World Commission on Environment and Development, *Our Common Future* (Oxford: Oxford University Press, 1987) (referred to as The Brundtland Report).
2. In Australia *The Helsham Report* provided an example of the dramatic political problems the environmental issue poses. It recommended that small areas of around 8 per cent of the area qualifying for a World Heritage listing should be set aside to preserve biological diversity and satisfy Australia's World Heritage obligations. The majority decision was that the remaining 92 per cent of the World Heritage listed area in Lemonthyme could be developed; The Helsham Report, *Report of the Commission of Inquiry into the Lemonthyme and Southern Forests*, Vols 1 and 2 (Canberra: AGPS, 1987).
3. Lars Samuelsson, 'On the Possibility of Evidence for Intrinsic Value in Nature', *Ethics and The Environment*, 18(2) (2013): 101–114 and for moral demands see Garrett Cullity, *The Moral Demands of Affluence* (Oxford: Clarendon Press, 2004). For early views on these global issues see Peter Singer, 'Famine, Affluence and Morality', *Philosophy and Public Affairs*, 1 (1971): 229–243.
4. I will use the capitalisation of the word 'Deep Ecology' as used by Bill McKibbon, *Deep Economy: Economics As If the World Mattered* (Oxford: One World, 2007).
5. Charles Taylor, 'Response to my Commentators', *Philosophy and Phenomenological Research*, 54(1) (1994): 213.
6. Charles Taylor, 'A Catholic Modernity?' In *Dilemmas and Connections* (Cambridge, MA: Harvard University Press, 2011): 167–187. Originally published as C. Taylor, 'A Catholic Modernity'. In *A Catholic Modernity?: Charles Taylor's Marianist Award Lecture*, edited by J. L. Heft (Oxford: Oxford University Press, 1999): 13–39.
7. Charles Taylor, *Sources of the Self* (Cambridge: Cambridge University Press, 1989): ch. 12.
8. Charles Taylor, 'Overcoming Epistemology'. In *Philosophical Arguments* (Cambridge, MA: Harvard University Press, 1995): 7. Charles Taylor, 'Overcoming Epistemology'. In *After Philosophy: End or Transformation?*, edited by Kenneth Baynes, James Bohman and Thomas McCarthy (Cambridge, MA: MIT Press, 1987): 464–488.
9. Bryan E. Bannon, 'From Intrinsic Value to Compassion: A Place-Based Ethic', *Environmental Ethics*, 35(3) (2013): 259–278.
10. Charles Taylor, 'Language Not Mysterious?' In *Reading Brandom: On Making It Explicit*, edited by Bernhard Weiss and Jeremy Wanderer (London: Routledge, 2010): 32–47.
11. Charles Taylor, 'Growth, Legitimacy and Modern Identity', *Praxis International* 1(2) (1981): 111. Reprinted as 'Legitimation Crisis?' [Growth, Legitimacy

and the Modern Identity] In *Philosophy and the Human Sciences: Philosophical Papers 2* (Cambridge: Cambridge University Press, 1985): 248–288.

12. Michael Mesner, 'What Might Environmental Philosophy Learn from Iris Young's Metropolitan Regionalism', *Environmental Ethics*, 35(4) (2013): 439–461.

13. Charles Taylor, 'How to Define Secularism'. In *Boundaries of Toleration*, edited by Charles Taylor and Alfred C. Stepan (New York: Columbia University Press, 2014): 59–78.

2 Basic Issues in Taylor's Philosophy

1. David Henry Thoreau, *Walking* (1862). Originally published in *The Atlantic Monthly*, 9(56): 657–674. Copy found at http://www.wildnesswithin. com/2001/01-12/inwildness.html (accessed 9 March 2014).

2. Charles Taylor, *The Ethics of Authenticity* (Oxford: Oxford University Press, 1992): 11–12.

3. Charles Taylor, 'Response to my Commentators', *Philosophy and Phenomenological Research*, 54(1) (1994): 213.

4. G. W. F. Hegel, *Philosophy of Nature*, edited and translated by A. V. Miller (foreword by J. N. Findlay) (Oxford: Clarendon Press, 1970).

5. See A. Ferrara, *Reflective Authenticity* (London: Routledge, 1998): 26–27.

6. Hubert Dreyfus, *Being and Time* (Cambridge, MA: MIT Press, 1994): 173–174.

7. Charles Taylor, 'Explanation and Practical Reason'. In *The Quality of Life*, edited by Martha Nussbaum and Amartya Sen (Oxford: Clarendon Press, 1983): 208–210.

8. Charles Taylor and Jürgen Habermas, 'Jürgen Habermas and Charles Taylor in Conversation', *The Immanent Frame*. http://blogs.ssrc.org/tif/2009/11/20/ rethinking-secularism-jurgen-habermas-and-charles-taylor-in-conversation/ 2008. And Charles Taylor and Jürgen Habermas, 'Dialogue'. In *The Power of Religion in the Public Sphere*, edited by Judith Butler (New York: Columbia University Press, 2011).

9. ibid., 173–174, 242–244.

10. Charles Taylor, 'Heidegger, Language, and Ecology'. In *Heidegger: A Critical Reader*, edited by H. L. Dreyfus and H. Hall (Oxford: Blackwells, 1992): 247–269 (reprinted in Charles Taylor, 'Heidegger, Language and Ecology'. In *Philosophical Arguments*, edited by Charles Taylor (Cambridge, MA: Harvard University Press, 1995): 100–126 and Charles Taylor, 'Language and Society'. In *Communicative Action*, edited by A. Honneth and H. Joas (Cambridge, MA: MIT Press, 1991): 23–36.

11. Taylor, *Ethics of Authenticity*, 7–10.

12. This interpretivist approach underlines Charles Taylor's response to an empiricist critique levelled by A. C. Grayling. Found in Ben Rogers, 'Charles Taylor Interviewed', *Prospect*, 29 February 2008, 4.

13. Russell Hittinger, 'Charles Taylor: *Sources of the Self*', *Review of Metaphysics*, 44(1) (1990): 111–130.

14. The term 'iron cage' is from Max Weber in his *The Protestant Ethic and the Spirit of Capitalism* (London: Alan and Unwin, 1965); B. S. Turner, *For Weber* (Boston, MA: Routledge and Kegan Paul, 1981). Weber explained how

modern bureaucracies perpetuate procedural structures that stifle the human condition.

15. Charles Taylor, 'Challenging Issues about the Secular Age', *Modern Theology*, 26(3) (2010): 404–416.
16. Charles Taylor, *Sources of the Self* (Cambridge: Cambridge University Press, 1989): 18–21. For a critical perspective on the theological dimension see Mark Redhead, 'Charles Taylor's Nietzschean Predicament: A Dilemma More Revealing than Foreboding', *Philosophy and Social Criticism*, 27(6) (2001): 81–107.
17. Hittinger, 'Charles Taylor: *Sources of the Self*', 111–130. See also J. Casanova, 'A Secular Age: Dawn or Twilight? In *Varieties of Secularism: In a Secular Age*, edited by M. Warner, J. Van Antwerpen and C. Calhoun (Harvard: Harvard University Press, 2010): 265–282.
18. ibid.
19. Taylor, 'Response to My Commentators', 213.
20. For example, Kant's categorical imperative binds all rational actors who utilise this faculty of instrumental conception of practical reasoning through time and space. According to Taylor, liberalism relies on instrumental and technical reasoning but this modern way of thinking about politics glides over other approaches to practical reasoning that concern what is the good which frames how nature is defined.
21. Charles Taylor, 'Politics of the Steady State', *New Universities Quarterly*, 32(2) (1978): 157–184.
22. Ilya Prigogine and Isabella Stengers, *Order Out of Chaos* (London: Fontana Books, 1984).
23. Isabella Stengers, *Cosmopolitics I* (Minneapolis: University of Minnesota and Press 2010) and Isabella Stengers, *Cosmopolitics II* (Minneapolis: University of Minnesota Press 2011).
24. Prigogine and Stengers, *Order Out of Chaos*, 312–313.
25. M. C. Watson, 'Derrida, Stengers, Latour, and Subalternist Cosmopolitics', *Theory, Culture and Society*, 31(1) (2014): 75–98.
26. I. Prigogine, *The End of Certainty* (New York: The Free Press, 1997): 66 (Initially found in B. Brugger and D. Kelly, *Chinese Marxism in the Post-Mao Era* (California: Stanford University Press, 1990), 56. Therefore, nature is but one of many systems that exhibit dissipative structures.
27. ibid., 56–57.
28. ibid.
29. Jim Thompson, 'A Refutation of Environmental Ethics', *Environmental Ethics*, 12(2) (1990): 147–160.
30. Stengers, *Cosmopolitics I* and Stengers, *Cosmopolitics II*, passim.
31. Taylor, 'Politics of the Steady State', 157–184.
32. Taylor, 'Response to my Commentators', 213.
33. Lars Samuelson, 'On the Possibility of Evidence for Intrinsic Value in Nature', *Ethics and the Environment*, 18(2) (2013): 111.
34. L. P. Hinchman, 'Aldo Leopold's Hermeneutic of Nature', *The Review of Politics*, 57(2) (1995): 225–249.
35. J. Baird Callicott, *In Defence of the Land Ethic: Essays in Environmental Philosophy* (Albany: State University of New York Press, 1989) and see J. Baird Callicott (2005a [1988]) 'Animal Rights and Environmental Ethics: Back Together

Again'. In *Environmental Philosophy: From Animal Rights to Radical Ecology*, 4th edition, edited by M. E. Zimmerman et al. (Upper Saddle River, NJ: Pearson Prentice-Hall): 130–138.

36. J. Baird Callicott, 'Hume's Is/Ought Dichotomy and the Relation of Ecology to Leopold's Land Ethic', *Environmental Ethics*, 4(2) (1982): 163–175 and for a detailed critique see Y. S. Lo, 'The Land Ethic and Callicott's Ethical System (1980–2001): An Overview and Critique', *Inquiry*, 44(2) (2001): 331–358.

37. See Bryan G. Norton, *Toward Unity among Environmentalists* (Oxford: Oxford University Press, 1991) and *Sustainability: A Philosophy of Adaptive Ecosystem Management* (Chicago: University of Chicago Press, 2005).

38. Bryan G. Norton, 'Epistemology and Environmental Values', *The Monist*, 75(2) (1992): 220. Norton is sufficiently postmodernist to argue that 'truth, as well as value, will be measured against a dynamic criterion of adaptability, rather than against a timeless, realist criterion of objective truth'.

39. ibid.

40. Charles Taylor, *A Secular Age* (Cambridge, MA: Harvard University Press, 2007): 338.

41. Charles Taylor, 'Foreword'. In *The Disenchantment of the World: A Political History of Religion*, edited by M. Cauchet (Princeton, NJ: Princeton University Press, 1997): x and initially found in Ruth Abbey, *Charles Taylor* (Oxford: Oxford University Press, 2001): 202.

42. ibid.

43. Charles Taylor, 'Modes of Civil Society', *Public Culture*, 3(1) (1990): 95–118.

44. See Taylor, *Sources of the Self*. In Chapter 7 'God Loveth Adverbs' Taylor connects these arguments with a model of practical reason and best account argument.

45. Jürgen Habermas, *The Structural Transformation of the Public Sphere: An Inquiry into a Category of Bourgeois Society*, translated by T. Burger (Cambridge, MA: MIT Press, 1992): 65–66. Charles Taylor, 'Modernity and the Rise of the Public Sphere'. In *The Tanner Lectures on Human Values, 14*, edited by Grethe B. Peterson (Salt Lake City: University of Utah Press, 1993): 203–260.

46. Taylor, Charles Taylor, *Modern Social Imaginaries* (Durham, NC: Duke University Press, 2004): 23.

47. Charles Taylor, 'On Social Imaginary', 2001, 23–24, internet version http://www.nyu.edu/classes/calhoun/Theory/Taylor-on-si.htm (and reproduced in Taylor, A Secular Age, p. 186).

48. Charles Taylor and Jürgen Habermas, 'Jürgen Habermas and Charles Taylor in Conversation', *The Immanent Frame*, 2009, http://blogs.ssrc.org/tif/2009/11/20/rethinking-secularism-jurgen-habermas-and-charles-taylor-in-conversation/.

49. Quoted in Habermas, *The Structural Transformation of the Public Sphere*, 65–66.

50. Taylor, *A Secular Age*, 189.

51. ibid., 189–193.

52. S. Gill, 'Toward a Postmodern Prince? The Battle in Seattle as a Moment in the New Politics of Globalisation', *Journal of International Studies*, 29(1) (2000): 131–140 and S. Gill, 'Globalisation, Market Civilisation, and Disciplinary Neo-Liberalism', *Millenium*, 24 (1995): 399–423.

3   Taylor's Interpretivism, Knowledge and the Natural Environment

1. Johann G. Herder, 'On the Origin of Language'. In *J. G. Herder on Social and Political Culture*, edited and translated by F. M. Barnard (Cambridge, MA: Cambridge University Press, 1969).
2. Hamann and others, *Was ist Aufklarung?*, edited by Ehrhard Bahr (Stuttgart: Philipp Reclam Verlag, 1974). Found in Frederick C. Beiser, *The Fate of Reason: German Philosophy from Kant to Fichte* (Cambridge, MA: Harvard University Press, 1987).
3. D. H. Thoreau, *Walking* (1862). Originally published in *The Atlantic Monthly*, 9(56): 657–674. Copy found at http://www.wildnesswithin.com/2001/01-12/inwildness.html (accessed 9 March 2014).
4. Charles Taylor, 'The Importance of Herder'. In *Isaiah Berlin: A Celebration*, edited by E. Margalit and A. Margalit (Chicago: University of Chicago Press, 1991: 40–64) (reprinted in Charles Taylor, *Philosophical Arguments* (Cambridge, MA: Harvard University Press, 1995): 79–100.
5. J. Baird Callicott, Johnathan Parker, Jordan Batson, Nathan Bell, Keith Brown, Samantha Moss, Alexandra Pool and John Wooding, 'The Other in a Sand County Almanac: Aldo Leopold's Animals and His Wild-Animal Ethic', *Environmental Ethics*, 33(2) (2011): 115–147.
6. Alasdair MacIntyre explains that the adjectives communitarian, communitive and communist were once used interchangeably, and that the term communist incorporated many diverse strands of thought until the early part of the 1920s; Alisdair MacIntyre, 'The Spectre of Communitarianism', *Radical Philosophy*, 70 (1995): 34–35. Some Deep Ecologists use the term communitarian in a different sense, referring to a 'biotic community'; see J. B. Callicott, 'Do Deconstructive Ecology and Sociobiology Undermine Leopold's Land Ethic', *Environmental Ethics*, 18(4) (1996): 353–373.
7. Philip Cafaro and Winthrop Staples III, 'The Environmental Argument for Reducing Immigration in the United States', *Environmental Ethics*, 31(1) (2009): 5–31 and see J. E. Cohen, 'How Many People Can the Earth Support?', *The New York Review of Books*, 45(15) (1998): 29–31.
8. Taylor now prefers the label 'post-liberal'. In this essay I use the term 'interpretivists' to refer to the ideas of Dreyfus, Gadamer, Nussbaum and Taylor. R. Abbey, 'Taylor as a Postliberal Theorist of Politics'. In *Perspectives on the Philosophy of Charles Taylor, Acta Philosophica Fennica*, edited by Arto Laitinen and Nicholas H. Smith (2002): 149–65 and R. Abbey, *Charles Taylor* (Oxford: Oxford University Press, 2001).
9. Charles Taylor, 'Interpretation and the Sciences of Man', *The Review of Metaphysics*, 25(1) (1971): 3–51.
10. Kenneth Baynes, 'Self, Narrative and Self-Constitution: Revisiting Taylor's "Self-Interpreting Animals"', *The Philosophical Forum*, 41(4) (2010): 441.
11. Charles Taylor, 'Interpretation and the Sciences of Man', *The Review of Metaphysics*, 25(1) (1971): 3.
12. Charles Taylor, *Hegel* (Cambridge: Cambridge University Press, 1975): 49–50.
13. ibid.

14. Martha Nussbaum, *Frontiers of Justice: Disability, Nationality and Species Membership* (Cambridge, MA: Belknap Press, Harvard University Press, 2007): 365.
15. From the letter to Gibieuf, 19 January 1642, which comes from the mediational passage 'You inquire about the principle by which I claim to know that the idea I have of something is not *an idea made inadequate by the abstraction of my intellect*. I derive this knowledge purely from my own thought or consciousness. I am certain that I can have no knowledge of what is outside me except by means of the ideas I have within me; and so I take great care not to relate my judgements immediately to things in the world, and not to attribute to such things anything positive which I do not first perceive in the ideas of them: *Descartes Philosophical Letters*, translated by Anthony Kenny (Oxford: Clarendon Press, 1970): 123.
16. Taylor as quoted in H. L. Dreyfus, 'Taylor's (Anti-) Epistemology'. In *Charles Taylor*, edited by R. Abbey (Cambridge: Cambridge University Press, 2004): 53.
17. Andrew Brennan, 'Moral Pluralism and the Environment', *Environmental Values* 1(1) (1992): 15–33 and J. Baird Callicott, *Companion to a Sand County Almanac: Interpretive and Critical Essays* (Madison: University of Wisconsin Press, 1987).
18. Charles Taylor, *Modern Social Imaginaries* (Durham, NC: Duke University Press, 2004): 2.
19. Ruth Abbey, *Philosophy Now: Charles Taylor* (Teddington and Princeton, NJ: Acumen Press/Princeton University Press 2000) and Ruth Abbey, 'Taylor as a Postliberal Theorist of Politics'. In *Perspectives on the Philosophy of Charles Taylor*, edited by A. Laitinen and N. Smith (*Acta Philosophica Fennica* 71, Helsinki: The Philosophical Society of Finland 2002): 149–161.
20. Charles Taylor, 'Overcoming Epistemology'. In *Philosophical Arguments* (Cambridge, MA: Harvard University Press, 1995): 7.
21. Charles Taylor, *Sources of the Self* (Cambridge: Cambridge University Press, 1989): 364.
22. ibid.
23. Taylor, 'Overcoming Epistemology', 8.
24. Charles Taylor, 'The Validity of Transcendental Arguments'. In *Philosophical Arguments*, edited by Charles Taylor (Cambridge, MA: Harvard University Press, 1995): 25.
25. Johnathan Blakely, 'Returning to the Interpretive Turn: Charles Taylor and His Critics', *The Review of Politics*, 75(3) (2013): 390.
26. ibid.
27. Charles Taylor, 'What Is Human Agency'. In *Philosophical Papers*, Vol. 1, edited by Charles Taylor (Cambridge: Cambridge University Press, 1985): 25–26 (Originally Charles Taylor, 'What Is Human Agency'. In *The Self: Psychological and Philosophical Issues*, edited by T. Mischel (Oxford: Oxford University Press, 1977): 114–115.
28. Baynes, 'Self, Narrative and Self-Constitution: Revisiting Taylor's "Self-Interpreting Animals"', 442.

4 Taylor's Interpretivism, Social Imaginaries and the Natural Environment

1. Charles Taylor, 'Contemporary Social Imagining', 2002, http://www.nyu.edu/classes/calhoun/theory/Taylor-on-si.htm; Charles Taylor, 'Modern Social

Imaginaries', *Public Culture* 14(1) (2002): 91–124. Charles Taylor, *Modern Social Imaginaries*, edited by Dilip Gaonkar, Jane Kramer, Benjamin Lee and Michael Warner (Durham, NC: Duke University Press, 2004).

2. Charles Taylor, 'On Social Imaginaries'. In *Traversing the Imaginary: Richard Kearney and the Postmodern Challenge*, edited by Peter Gratton and John Panteleimon Manoussakis (Evanston: Northwestern University Press, 2007): 29–48

3. Charles Taylor, 'Politics of the Steady State', *New Universities Quarterly* 32(2) (1978): 157–184 and see Taylor, 'How to Define Secularism', 59–78.

4. Hans Georg Gadamer, *Truth and Method*, Fifth reproduction, translated by Garrett Barden and John Cummings (London: Sheed and Ward Ltd, 1975) and also H. G. Gadamer, *Hegel's Dialectic: Five Hermeneutical Studies* (New Haven, CT: Yale University Press, 1976).

5. Martin Heidegger, *Being and Time*, translated by J. Macquarie and Edward Robinson (New York: Harper and Rowe, 1962).

6. Martha Nussbaum, *Frontiers of Justice: Disability, Nationality and Species Membership* (Cambridge, MA: Belknap Press, Harvard University Press, 2007): 365.

7. Charles Taylor, 'Heidegger, Language, and Ecology'. In *Heidegger: A Critical Reader*, edited by H. L. Dreyfus and H. Hall (Oxford: Blackwells, 1992): 247–269.

8. Charles Taylor, *Modern Social Imaginaries* (Durham, NC: Duke University Press, 2004): 23.

9. Taylor, 'Modern Social Imaginaries', 91.

10. Taylor makes this point in his lecture – 'Can Human Action Be Explained' – http://vimeo.com/7803207 accessed 1 January, 2015.

11. Charles Taylor, *Sources of the Self* (Cambridge: Cambridge University Press, 1989): ch. 12.

12. Charles Taylor, 'Foundationalism and the Inner-Outer Distinction'. In *Reading McDowell: On Mind and World*, edited by N. H. Smith (London: Routledge, 2002): 106–121.

13. ibid., 106–121.

14. Charles Taylor, 'Afterword: Apologia pro Libro suo'. In *Varieties of Secularism in a Secular Age*, edited by Michael Warner, Jonathan VanAntwerpen and Craig J. Calhoun (Harvard: Harvard University Press, 2010): 302–303.

15. Charles Taylor, 'Buffered and Porous Selves', *The Immanent Frame*, 2 September 2008, http://blogs.ssrc.org/tif/2008/09/02/buffered-and-porous-selves.

16. ibid.

17. ibid.

18. Charles Taylor, 'Recovering the Sacred', *Inquiry* 54(2) (2011): 117–118.

19. For example, the economist Julian Simon is notable for his supposition that the concept of an environmental crisis is a fiction. While this type of extreme economic thinking is rarely endorsed these days, there still has only been limited research on the social imaginaries that challenge the hegemony of this economic logic.

20. W. G. F. Hegel, *Early Theological Writings*, edited by T. M. Knox (Chicago: University of Chicago Press, 1948).

21. J. Baird Callicott, Johnathan Parker, Jordan Batson, Nathan Bell, Keith Brown, Samantha Moss, Alexandria Pool and John Wooding, 'The Other in a

Sand County Almanac: Aldo Leopold's Animals and His Wild-Animal Ethic', *Environmental Ethics*, 33(2) (2011): 115–147.
22. Iain Thomson, 'Transcendence and the Problem of Otherworldly Nihilism: Taylor, Heidegger, Nietzsche', *Inquiry*, 54(2) (2011): 144.
23. Charles Taylor, *A Secular Age* (Cambridge, MA: Harvard University Press, 2007): 14.
24. Stephen K. White, 'Fullness and Dearth: Depth Experience and Democratic Life', *American Political Science Review*, 104(4) (2010): 801.
25. ibid.
26. ibid., 815.
27. John Rundell, 'Charles Taylor's Search for Transcendence: Mystery, Suffering, Violence'. In *Secularisations and Their Debates, Perspectives on the Return of Religion in the Contemporary West*, edited by M. Sharpe and D. Nickelsen (Dordrecht: Springer, 2014): 211.
28. ibid.
29. ibid., 211–212.
30. ibid., 211.
31. ibid. 'Here mystery, in this sense, is polytheistic, rather than in a consumerist sense.'
32. ibid. As Rundell points out this is defended in Taylor's work on the specificity of *Quebequois* culture.
33. ibid.
34. ibid.
35. Jürgen Habermas, '"The Political": The Rational Meaning of a Questionable Inheritance of Political Theology'. In *The Power of Religion in the Public Sphere*, edited by Judith Butler, Jürgen Habermas, Charles Taylor and Cornell West (New York: Columbia University Press, 2011).

5 Taylor's Metaphysics, Merleau-Ponty and the Natural Environment

1. The Enlightenment emphasised a new age of science and technology to overcome myth and superstition – this epoch ushered in the view that humanity could control the natural environment. For environmental purposes, Taylor returns to the work of Hegel's G. W. F. Hegel, *Philosophy of Nature*, edited and translated by A. V. Miller (foreword by J. N. Findlay) (Oxford: Clarendon Press, 1970).
2. J. Baird Callicott, 'Humes Is/Ought Dichotomy and the Relation of Ecology to Leopold's Land Ethic', *Environmental Ethics*, 4(2) (1982): 163–175.
3. Aldo Leopold, *A Sand Country Almanac* (New York: Sierra Club/Ballantine, 1970).
4. J. Baird Callicott, 'Rolstrom on Intrinsic Value: A Deconstruction', *Environmental Ethics*, 14(2) (1992):129–145.
5. In this chapter I extend interpretivism to represent a particular approach that is critical of contemporary Cartesian and Humean approaches to the formation of knowledge, I also use the natural environment to explore the cosmological imaginary that guides Taylor's search for external reasons and values.
6. This is in the public sphere where discussions about inherent, intrinsic and normative claims are given consideration.
7. Charles Taylor, 'Overcoming Epistomology'. In Philosophical Arguments (Cambridge, MA: Harvard University Press, 1995): 1. Charles Taylor, 'The

Validity of Transcendental Arguments'. In *Philosophical Arguments*, edited by Charles Taylor (Cambridge, MA: Harvard University Press, 1995): 23–25.
8. Taylor, 'The Validity of Transcendental Arguments', 25. And *Proceedings of the Aristotelian Society*, New Series, Vol. 79 (1978–79): 153.
9. Taylor, 'The Validity of Transcendental Arguments', 153.
10. In earlier work Taylor focuses on Descartes' work, initiating the inner representational view of understanding.
11. In other work Taylor goes back to Descartes to explain that political theory relies excessively on adapting a model of the natural sciences without full consideration of mind and world dualisms which has led to a (mis)understanding of what it is to know, which has had dire effects on both theory and practice in a host of domains. Taylor is criticising Descartes' mediational picture of the world and offers a different theory that the mind is in the world and not separate from it. He argues that Descartes picture was part of a Scientific Revolution that had a predilection for measurement, procedure and technique. This framework has been applied to the humanities without fully addressing its applicability – Taylor's concern is that scientific approaches to governance are not the means to order human differences and affairs.
12. Taylor, Validity of Transcendental Arguments', 154.
13. Charles Taylor, 'Overcoming Epistemology'. In *Philosophical Arguments* (Cambridge, MA: Harvard University Press, 1995): 1.
14. Charles Taylor, 'What Is Secularity'. In *Transcending Boundaries in Philosophy and Theology: Reason, Meaning and Experience*, edited by K. Vanhoozer and Martin Warner (Burlington, VT: Ashgate, 2007): 60.
15. ibid.
16. ibid., 61.
17. John McDowell, *Mind and World* (Cambridge, MA: Harvard University Press, 1994): 85.
18. Hubert L. Dreyfus, 'Taylor's Anti-Realism'. In *Charles Taylor*, edited by R. Abbey (Cambridge: Cambridge University Press, 2004): 66. Dreyfus was quoting a private communication with Taylor on anti-epistemology.
19. Callicott et al. 'The Other in a Sand County Almanac', 144.
20. Leopold, 'A Sand County Almanac', vii.
21. Taylor, *A Secular Age…*, 681.
22. John Rundell, 'Charles Taylor's Search for Transcendence: Mystery, Suffering, Violence'. In *Secularisations and Their Debates, Perspectives on the Return of Religion in the Contemporary West*, edited by M. Sharpe and D. Nickelsen (Dordrecht: Springer, 2014).
23. ibid.
24. Taylor, 'Overcoming Epistemology', 474.
25. Charles Taylor, *Hegel* (Cambridge: Cambridge University Press, 1975): 19.
26. ibid., 30–42.
27. Herder, *Ideas for the Philosophy of History*, found in Forster, 'Herder's Philosophy of Language', 324 (located in G6: 347; *Metacritique* S21:19, 88, 293–4).
28. Charles Taylor, 'Heidegger, Language, and Ecology'. In *Heidegger: A Critical Reader*, edited by H. L. Dreyfus and H. Hall (Oxford: Blackwells, 1992): 247–269.
29. Charles Taylor, 'Recovering the Sacred', *Inquiry* 54(2) (2011): 117–118.
30. ibid., 117–118.
31. Kenneth Baynes, 'Self, Narrative and Self-Constitution: Revisiting Taylor's "Self-Interpreting Animals"', *The Philosophical Forum*, 41(4) (2010): 441.

32. Taylor, 'Recovering the Sacred', 117.
33. Bryan G. Norton, 'Epistemology and Environmental Values', *The Monist*, 75(2) (1992): 224 and for Callicott's response to this line of thinking see 'Was Aldo Leopold a Pragmatist? Rescuing Leopold from the Imagination of Bryan Norton' (with William Grove-Fanning, Jennifer Rowland, Daniel Baskind, Robert Heath French, and Kerry Walker) *Environmental Values* 18 (2009): 453–486.
34. Baynes, 'Self, Narrative and Self-Constitution', 442.
35. See Taylor, 'Heidegger, Language and Ecology', 266–269.
36. ibid.
37. Jim Cheney, 'Beyond Subjectivism and Objectivism', *The Monist*, 75(2) (1992): 233.
38. ibid., 24.

6 Taylor's Environmentalism and Critique of Utilitarianism and Instrumental Reason

1. It is interesting to note that modern environmental consequentialists observe that there have been limited arguments using the doctrine in environmental ethics. See Avram Hiller and Leonard Kahn, 'Introduction: Consequentialism and Environmental Ethics'. In *Consequentialism and Environmental Ethics*, edited by Avram Hiller, Ramona Ilea and Leonard Kahn, Routledge Studies in Ethics and Moral Theory (New York: Routledge, 2014).
2. Alan Carter, 'Indirect, Multidimensional Consequentialism'. In *Consequentialism and Environmental Ethics*, edited by Avram Hiller, Ramona Ilea and Leonard Kahn, Routledge Studies in Ethics and Moral Theory (New York: Routledge, 2014): 70–92.
3. See Derek Parfit, 'On Doing the Best for Our Children'. In *Ethics and Population*, edited by M. D. Bayles (Cambridge, MA: Schenkman Publishing Company Inc., 1976): 100–115; 'Future Generations: Further Problems', *Philosophy and Public Affairs*, 11 (1982): 113–172; *Reasons and Persons* (Oxford: Clarendon Press; 1986); 'Overpopulation and the Quality of Life'. In *Applied Ethics*, edited by P. Singer (Oxford: Oxford University Press): 145–164 (Reprinted in J. Ryberg and T. Tännsjö (eds.), 'Postscript'. In *The Repugnant Conclusion. Essays on Population Ethics*, edited by J. Ryberg and T. Tännsjö (Dordrecht: Kluwer Academic Publishers, 2004): 257.
4. Charles Taylor, *Sources of the Self* (Cambridge: Cambridge University Press, 1989): 347.
5. ibid., 80.
6. The most famous polemic piece on this subject is Alasdair MacIntyre's *After Virtue* (London: Duckworth, 1981): 6–12 which provides perhaps the most notable critique of the political institutions of modernity in recent times.
7. Immanuel Kant, *Groundwork on the Metaphysics of Morals* (originally *The Moral Law*) translated by H. J. Paton (London: Hutchinson, 1949); Taylor, *Sources of the Self*, 70–80.
8. Alasdair MacIntyre, *A Short History of Ethics* (New York: Collier Books, 1966): 1–3.
9. Brian G. Wolff, 'Environmental Studies and Utilitarian Ethics', *Environmental Studies* 34(2) (2008): 6–11.

10. MacIntyre, *After Virtue*, 156.
11. Charles Taylor, 'Alternative Futures: Legitimacy, Identity and Alienation in Late-Twentieth Century Canada'. In *Constitutionalism, Citizenship and Society in Canada*, edited by A. Cairns and C. Williams (Toronto: University of Toronto Press, 1985): 183–229 [Reprinted in C. Taylor, 'Alternative Futures'. In *Reconciling the Solitudes*, edited by G. Laforest (Montreal: McGill Queen's University Press, 1993): 59–120.
12. R. Garner, *A Theory of Justice for Animals: Animal Rights in a Nonideal World* (Oxford: Oxford University Press, 2013).
13. R. G. Collingwood, *The Idea of Nature* (Oxford: Oxford University Press, 1945): 7; initially found in L. P. Hinchman, 'Aldo Leopold's Hermeneutic of Nature', *The Review of Politics*, 57(2) (1994): 225–249.
14. John Stuart Mill, 'Nature'. In *Nature, the Utility of Religion and Theism* (London: Longmans, Green Reader, and Dyer, MDCCCLXXIV): 3–69. Mill's instrumental perspective seemed to support unlimited economic growth though in his *Principles of Political Economy* he deplored conspicuous consumption and interestingly championed a stationary-state; John Stuart Mill, *Principles of Political Economy* (New York: Longmans, Green and Company, 1929): 748–750.
15. Mill, 'Nature', 20–21.
16. Though it is worth remembering that a lot of Mill's work actually had a romantic view of nature.
17. M. Lewis, *Green Delusions: An Environmentalist Critique of Radical Environmentalism* (Durham, NC: Duke University Press, 1992).
18. Peter Singer, *Animal Liberation* (London: Jonathan Cape, 1976).
19. Robert E. Goodin, *Green Political Theory* (Massachusetts: Polity Press, 1992).
20. Tom Regan, *The Case for Animal Rights* (Berkeley: University of California Press, 1983).
21. See for example P. W. Taylor, *Respect for Nature: A Theory of Environmental Ethics* (Princeton: Princeton University Press, 1986).
22. Charles Taylor, 'Atomism'. In *Philosophy and the Human Sciences. Philosophical Papers II*, edited by Charles Taylor (Cambridge: Cambridge University Press, 1985): 187–210.
23. L.E. Johnson, *A Morally Deep World* (Cambridge: Cambridge University Press, 1991), 205
24. ibid., 282
25. ibid., 287 (Johnson explains the notion of moral considerability: The thing for us to do is to find our way in the world, while giving due respect to the widely disparate interests of other beings. This is not essentially different from what we ought to do concerning other humans. Other humans also have widely divergent interests, ranging from the insignificant to the monumental, and it is for us to make the appropriate moral responses. In morally dealing with others, human or nonhuman, we must estimate the importance of various interests of various entities (including ourselves). In so doing it would be appropriate to consider how highly developed an entity and its interests are, how vital an interest is to it, and what the alternatives are, 287
26. See Mill, 'Nature', 18–19.
27. See Robert E. Goodin, *Utilitarianism as a Public Philosophy* (Cambridge: Cambridge University Press, 1995): 48.
28. ibid., 40.

29. ibid.
30. Robyn Eckersley, 'Just Natural Relations? Recent Developments in Environmental Theory', *Political Theory Newsletter*, 5(2) (1993): 118.
31. Robert Goodin, 'Utilitarianism as a Public Philosophy'. In *Political Theory*, edited by A. Vincent (Cambridge: Cambridge University Press, 1997): 67–89.
32. ibid.
33. Mark Sagoff, 'Ethics and Economics in Environmental Law'. In *Earthbound: New Introductory Essays in Environmental Ethics*, edited by T. Regan (New York: Random House, 1984): 175.
34. J. B. Callicott with Michael E. Zimmerman, Irene J. Klaver, Karen J. Warren, and John Clark, *Environmental Philosophy: From Animal Rights to Social Ecology*, Fourth edition (Upper Saddle River, NJ: Prentice-Hall, 2005).
35. Tom Regan, *The Case for Animal Rights* (Berkeley, CA: University of California Press, 1983).
36. See for example Paul. W. Taylor, *Respect for Nature: A Theory of Environmental Ethics* (Princeton, NJ: Princeton University Press, 1986) and also Wolff, 'Environmental Studies and Utilitarian Ethics', 6–11.
37. Charles Taylor, 'Atomism'. In *Philosophy and the Human Sciences: Philosophical Papers II*, edited by Charles Taylor (Cambridge: Cambridge University Press, 1985): 187–210.
38. Mark Sagoff, 'Ethics and Economics in Environmental Law'. In *Earthbound: New Introductory Essays in Environmental Ethics*, edited by Regan, T. (New York, Random House, 1984): 175.
39. ibid.
40. Taylor, *Sources*, 340·
41. Charles Taylor, 'Explanation and Practical Reason'. In *The Quality of Life*, edited by Martha Nussbaum and Amartya Sen (Oxford: Clarendon Press, 1983): 208–241. Taylor points out in his 'Response to Commentators' (discussed in Chapter 2) that it was a strategic error to have left out an explicit account of the relationship between *ad hominem* and *apodictic* conceptions of practical reasoning.
42. Taylor, 'Alternative Futures', 79. This argument is similar to that of Rosenblum who maintains that 'incessant willing [which] is strongest in utilitarianism, which unleashes desires, proclaims the sovereignty of pleasure and pain, and proposes maximising satisfaction as the sole imperative': Nancy Rosenblum, *Another Liberalism* (Cambridge, MA: Harvard University Press, 1987): 29.
43. John Stuart Mill, *Utilitarianism* (Indianapolis: Hackett Edition, 1979): 4–5, 34.
44. Taylor, 'Explanation and Practical Reason', 210.
45. ibid.
46. ibid.
47. See Will Kymlicka, 'The Ethics of Inarticulacy', *Inquiry* 34 (2) (1991): 155–182, 167.
48. Taylor, 'Alternative Futures', 79 and Taylor, *Sources of the Self*, 364.
49. Derek Parfit, 'Overpopulation and the Quality of Life'. In *Applied Ethics*, edited by P. Singer (Oxford: Oxford University Press, 1986): 145–165.
50. ibid·
51. ibid., 161 and N. M. Lemos, 'Higher Goods and the Myth of Tithonus', *Journal of Philosophy*, 90(9) (1993): 490.
52. Parfit, 'Overpopulation and the Quality of Life', 150.

53. He investigates Parfit's dilemma; Lemos, 'Higher Goods and the Myth of Tithonus', 482–496.
54. ibid., 487–488.
55. M. J. Zimmerman, 'Virtual Intrinsic Value and the Principle of Organic Unities', *Philosophy and Phenomenological Research*, 59(3) (1999): 654–667.
56. For a more detailed discussion concerning Mill's modernist environmental views see Chapter 4 of this book which uses Hinchman, 'Aldo Leopold's Hermeneutic of Nature', 225–249.
57. Aldo Leopold, *A Sand Country Almanac* (New York: Sierra Club/Ballantine, 1970).
58. ibid., 262.
59. See Hinchman, 'Aldo Leopold's Hermeneutic of Nature', 222.
60. Taylor, *Sources*, 359.
61. See K. Shrader-Frechette, 'Practical Ecology and Foundations for Environmental Ethics', *Journal of Philosophy*, 92(12) (1995): 621–635.
62. Taylor takes us back to romantic critics of the Enlightenment such as Hamann and Herder. Of course, as Charles Larmore points out many romantic frameworks exist and, in fact, some romantics preferred a form of Promethean individualism. See Charles Larmore, 'Political Liberalism', *Political Theory*, 18(3) (1990): 344. Although no clear definition of romanticism can be advanced the Romantic Movement provides a means to think about values submerged by Enlightenment rationality.
63. Rawls distances himself from Ronald Dworkin's interpretation of his theory as a version of (natural) right by stressing the role of intuition: J. Rawls, 'Justice as Fairness: Political not Metaphysical', *Philosophy and Public Affairs*, 14(3) (1985): 236n. This was considered in Chapter 2 of this book.
64. Charles Taylor, *The Ethics of Authenticity* (Oxford: Oxford University Press, 1992): 26.
65. Taylor, *Hegel*, 49–50.
66. Robert Goodin, a utilitarian philosopher, is prepared to take such a step. Goodin explains that value is relational in the sense that it presupposes a human valuing subject, but the natural value people feel when they confront nature is not simply a construction of humanity. Nature must possess value for people to feel when they come in contact with it (see Goodin, *Green Political Theory* ...). This has been developed by Wolff in his 'Environmental Studies and Utilitarian Ethics', 6–11.
67. Honneth argues that only communities which regard themselves as 'democratic constitutional states currently provide a sufficient guarantee that conflicts will be settled in a peaceful manner'; A. Honneth, 'Is Universalism a Moral Trap'. In *Perpetual Peace: Essays on Kant's Cosmopolitan Idea*, edited by J. Bohman and M. Lutz-Bachmann (Cambridge, MA: MIT Press, 1997): 176. Taylor offers a republican framework of a liberal order through which recognition of common relationships is developed through democratic means. See Taylor, 'Cross-purposes: The Liberal-Communitarian Debate', 159–182.
68. Taylor, 'Atomism', 191.
69. ibid., 191–192.
70. ibid., 192.
71. ibid.
72. ibid.

73. Charles Taylor, 'Responsibility for Self', *The Identities of Persons*, edited by A. O. Rorty (Berkeley, CA: University of California Press, 1969): 281–301.
74. Taylor, *Hegel*, 547.
75. Kymlicka criticises communitarians for treating *all* liberals as if they adhered to an atomist conception of the human agent: Kymlicka, *Liberalism, Community and Culture*, 47–48.
76. Hinchman, 'Aldo Leopold's Hermeneutic of Nature', 225–249.
77. Chapters 1 and 2 explore Taylor's hermeneutics.
78. Charles Taylor, *Hegel* (Cambridge: Cambridge University Press, 1975): 49–50.
79. ibid., 49–50.
80. C. P. Snow, *Two Cultures: And a Second Look* (Cambridge: Cambridge University Press, 1965).
81. Charles Taylor, 'Rationality'. In *Rationality and Relativism*, edited by M. Hollis and S. Lukes (Oxford: Basil Blackwell, 1981): 103.
82. See Charles Taylor, 'Understanding in Human Sciences', *Review of Metaphysics*, 34(1) (1980): 25–38.
83. Hubert Dreyfus, 'Holism and Hermeneutics', *Review of Metaphysics*, 34(1) (1980): 16–17.
84. Charles Taylor, interviewed by Ben Rogers, *Prospect Magazine* (29 February 2008): 16.
85. Charles Taylor, 'Responsibility for Self'. In *The Identities of Persons*, edited by A. O. Rorty (Berkeley, CA: University of California Press, 1969): 281–301.
86. Charles Taylor, 'What Is Human Agency'. In *Philosophical Papers*, Vol. 1, edited by Charles Taylor (Cambridge: Cambridge University Press, 1985): 25–26.
87. See Charles Taylor, 'The Dynamics of Democratic Exclusion', *Journal of Democracy*, 9(4) (1998): 143–156.
88. Callicott, 'Do Deconstructive Ecology and Sociobiology Undermine Leopold's Land Ethic?', 353–372. Callicott explains that the 'environmental crisis' is not a misguided 'social construct' created from a particular horizon of intelligibility, but evaluates the rate of change of biological diversity and the anthropogenic effects of humanity that substitutes domestic species for wild ones.

7 Taylor's Critique of Instrumentalism, Liberalism and Procedure in Politics

1. Rawls' liberalism has been adapted to nature in Bryan G. Norton and B. Hannon 'Environmental Values: A Place-Based Approach', *Environmental Ethics*, 19(3) (1997): 227–247; Ernest Partridge, 'Nature as a Moral Resource', *Environmental Ethics*, 6(2) (1984): 101–131; Christina Hoff, 'Kant's Invidious Humanism', *Environmental Ethics*, 5(1) (1983): 63–71; D. VanDeVeer, 'On Beasts, Persons, and the Original Position', *The Monist*, 62(3): 368–377. For Rawls' explanation of how his political liberal principles can be adapted to incorporate environmental concerns, see for example J. Rawls, *Political Liberalism* (New York: Columbia University Press, 1993): xxviii–xiv.
2. R. Abbey, 'Rawlsian Resources for Animal Ethics', *Ethics and the Environment*, 12(1) (2007): 1–28.
3. Charles Taylor, 'Modern Moral Rationalism'. In *Weakening Philosophy: Essays in Hr of Gianni Vattimo*, edited by Santiago Zabala (Montreal: McGill-Queen's University Press, 2007): 64.

4. Rawls, *Political Liberalism*, xxviii.
5. Rawls' liberalism has been adapted to nature in Bryan G. Norton and B. Hannon, 'Environmental Values: A Place-Based Approach', *Environmental Ethics*, 19(3) (1997): 227–247; E. Partridge, 'Nature as a Moral Resource', *Environmental Ethics*, 6(2) (1984): 101–131; Christina Hoff, 'Kant's Invidious Humanism', *Environmental Ethics*, 5(1) (1983): 63–71; D. VanDeVeer, 'On Beasts, Persons, and the Original Position', *The Monist*, 62(3) (XXXX): 368–377.). For Rawls' explanation of how his political liberal principles can be adapted to incorporate environmental concerns, see Rawls, *Political Liberalism*, xxviii–xiv.
6. Derek Bell, 'How Can Political Liberals Be Environmentalists?', *Political Studies*, 50(5) (2002): 703–724.
7. Will Kymlicka, *Liberalism, Community and Culture* (Oxford: Clarendon Press, 1989).
8. J. M. Meyer, 'We Have Never Been Liberal: The Environmentalist Turn to Liberalism and the Possibilities for Social Criticism', *Environmental Politics*, 20(3) (2011): 56–373.
9. Michael Ignatieff, *The Warrior's Honor* (London: Chatto and Windus, 1998): 66.
10. J. Rawls, *A Theory of Justice* (Oxford, Oxford University Press, 1971), 11.
11. Kant, *Groundwork on the Metaphysics of Morals*, 96.
12. Rawls, *Political Liberalism*, xxviii–xxiv.
13. Rawls' *Theory of Justice* expressed dissatisfaction with the tendency in modern political theory to rely on utilitarian political ideas. Rawls, in the tradition of the New Deal, clearly felt that utilitarianism neglected the poor and needy.
14. Rawls, 'Justice as Fairness: Political Not Metaphysical', 236n.
15. Rawls stated that political liberalism is committed to toleration, and he promised to apply 'the principle of toleration to philosophy itself' (Rawls, *Political Liberalism*, 10). Both Larmore and Rawls recognised the importance of culture and identity and argued that political liberalism was *strictly* political in the way it respected the diversity of different cultural, social and political ways of being.
16. Rawls, *Political Liberalism*, 181, provides a list of the primary goods for a Western liberal society. In the footnotes he pointed out that the list of primary goods could, conceivably, incorporate leisure time and other social goods.
17. Rawls, *A Theory*, 11.
18. See S. A. Schwarzenbach, 'Rawls, Hegel and Communitarianism', *Political Theory*, 19(4) (1991): 539–572.
19. Bell, 'How Can Political Liberals Be Environmentalists?', 705.
20. This is an example of one of the assorted inconsistencies in the doctrine, which have brought it ever increasing scrutiny. Many critics subsequently pointed out that not only were there problems with the derivation of his principles, but that the environment and its importance for humanity were being largely overlooked.
21. Rawls, *Political Liberalism*, 245.
22. John Rawls, *Justice as Fairness: A Restatement* (Cambridge, MA: Belknap Press, 2001): 152, note 2.

23. Philippe Van Parijs, 'Why Surfers Should Be Fed: The Liberal Case for an Unconditional Basic Income', *Philosophy and Public Affairs*, 20(2) (1991): 105.
24. Charles Taylor, 'Responsibility for Self'. In *The Identities of Persons*, edited by A. O. Rorty (Berkeley, CA: University of California Press, 1969): 296.
25. Charles Taylor, 'Can Liberalism Be Communitarian?', *Critical Review*, 8(2) (1994): 259–260.
26. ibid., 257–263.
27. Charles Taylor, *Sources of the Self*, 69.
28. The key difference we are looking at between our two marker dates is a shift in the understanding of what I call 'fullness' between a condition in which our highest spiritual and moral aspiration point us inescapably to God, one might say, make(s) no sense without God, to one in which they can be related to a host of different sources, and frequently are referred to sources which deny God (Charles Taylor, *A Secular Age* (Cambridge, MA: Harvard University Press, 2007): 26).
29. Taylor, *A Secular Age*, 26.
30. Charles Taylor, *The Meaning of Secularism*, 'The Meaning of Secularism' *The Hedgehog Review*, 12.3 (2010): 23–34. http://www.iasc-culture.org/HHR_Archives/Fall2010/Taylor_lo.pdf.
31. John Dunn, 'Capitalist Democracy: Elective Affinity or Beguiling Illusion?' *Daedalus*, 7, 136(3) (2007): 10.
32. Charles Taylor, *The Ethics of Authenticity* (Oxford: Oxford University Press, 1992): 6–11.
33. John Dunn, 'Capitalist Democracy: Elective Affinity or Beguiling Illusion?' *Daedalus*, 7, 136(3) (2007): 12.
34. The criterion of equal facilitation, advanced in the discourse arena, acts as an adjudicating mechanism among different claims. In endeavouring to find agreement among people, this mechanism performs a task similar to Rawls' principle of respect.
35. Charles Taylor, 'Explanation and Practical Reason'. In *The Quality of Life*, edited by M. Nussbaum and A. Sen (Oxford: Clarendon Press, 1993): 230.
36. ibid.
37. A similar interpretation of Hegel's legacy can be found in the work of Findlay and Hyppolite. See J. Hyppolite, *Genesis and Structure in Hegel's Phenomenology* (Evanston: Northwestern University Press, 1974) and J. Findlay, *Hegel: A Re-examination* (England: Greg Revivals, 1953): 36, 45.

8   Interpretation, Language and Environmental Values: The Habermas and Taylor Debate

1. Charles Taylor, 'Language Not Mysterious?' In *Reading Brandom: On Making It Explicit*, edited by Bernhard Weiss and Jeremy Wanderer (London: Routledge, 2010): 32–47.
2. Charles Taylor, 'The Politics of the Steady State'. In *Beyond Industrial Growth*, edited by Abraham Rotstein (Toronto: University of Toronto Press, 1976): 47–70. Also available as AV reel (Toronto: CBC Learning Systems, 1974–75).
3. Charles Taylor, 'Heidegger, Language, and Ecology'. In *Heidegger: A Critical Reader*, edited by H. L. Dreyfus and H. Hall (Oxford: Blackwells, 1992): 286.

4. Jürgen Habermas, '"The Political": The Rational Meaning of a Questionable Inheritance of Political Theology'. In *The Power of Religion in the Public Sphere*, edited by Judith Butler, Jürgen Habermas, Charles Taylor and Cornell West (New York: Columbia University Press, 2011): 1–15.

5. Habermas, J., 'Struggles for Recognition in the Constitutional Democratic State', in *Multiculturalism: Examining the Politics of Recognition*, Edited and Introduced by A. Gutmann, (ed.), (expanded edition of *Multiculturalism: Examining the Politics of Recognition*, 1992), Princeton, New Jersey: Princeton University Press, 1994): 130.

6. John Dryzek, 'Green Reason: Communicative Ethics for the Biosphere', *Environmental Ethics* 12(3) (1988): 210.

7. Charles Taylor, 'On Disclosing New Worlds', *Inquiry* 38(1/2) (1995): 119–123; Charles Taylor, 'The Importance of Herder'. In *Isaiah Berlin: A Celebration*, edited by E. Margalit and A. Margalit (Chicago: University of Chicago Press, 1991): 40–64 (reprinted in Charles Taylor, *Philosophical Arguments* (Cambridge, MA: Harvard University Press, 1995): 79–100). See also M. N. Forster, 'Herder's Philosophy of Language, Interpretation, and Translation: Three Fundamental Principles', *The Review of Metaphysics* 56(2) (2002): 323–354.

8. Taylor has also argued that the vision of a procedural world assumes that the natural environment simply stands in wait for people to use and manipulate. Taylor, 'Heidegger, Language, and Ecology', 247–269.

9. Charles Taylor, *The Ethics of Authenticity* (Oxford: Oxford University Press, 1992).

10. For Taylor's comments on Habermas' procedural *modus vivendi* see Charles Taylor, 'Ethics and Ontology', *Journal of Philosophy*, C(6) (2003): 305.

11. C. Taylor, 'Civil Society in the Western Tradition'. In *The Notion of Tolerance and Human Rights*, edited by E. Groffier and M. Paradis (Ottawa: Carleton University Press, 1991).

12. Charles Taylor, 'Alternative Futures: Legitimacy, Identity and Alienation in Late-Twentieth Century Canada'. In *Constitutionalism, Citizenship and Society in Canada*, edited by A. Cairns and C. Williams (Toronto: University of Toronto Press, 1985): 183–229 (reprinted in C. Taylor, 'Alternative Futures'. In *Reconciling the Solitudes*, edited by G. Laforest (Montreal: McGill, Queen's University Press 1993): 59–120.

13. Charles Taylor, 'A Tension in Modern Democracy'. In *Democracy and Vision*, edited by A. Botwinick and W. E. Connolly (Princeton and Oxford: Princeton University Press, 2001): 79–99.

14. Charles Taylor, 'What's Wrong with Foundationalism?: Knowledge, Agency and World'. In *Heidegger, Coping and Cognitive Science*, edited by M. A. Wrathall and J. Malpas (Massachusetts: University of Massachusetts Press, 2000): 115–134.

15. Taylor, 'On Disclosing New Worlds', 119–123.

16. Charles Taylor, *Hegel and Modern Society* (Cambridge: Cambridge University Press, 1979).

17. Jürgen Habermas, 'A Reply'. In *Communicative Action*, edited by A. Honneth and H. Joas (Cambridge, MA: MIT Press, 1991): 2216–217. See Taylor, 'Ethics and Ontology', 305–320.

18. Jürgen Habermas, *Justification and Application* (Cambridge, MA: MIT Press, 1993): 56–57 where he states that 'we presuppose a dialogical situation that

satisfies ideal conditions in a number of respects, including...freedom of access, equal rights to participate, truthfulness on the part of participants, absence of coercion in taking positions, and so forth' (56). Found initially in N. Porter, *Rethinking Unionism, An Alternate Vision for Northern Ireland* (Belfast: The Blackstaff Press, 1985).

19. See Charles Taylor, 'Language and Society'. In *Communicative Action*, edited by A. Honneth and H. Joas (Cambridge, MA: MIT Press, 1991): 23–25.
20. Habermas, 'A Reply', 216–217.
21. ibid.
22. J. Habermas, *Theory of Communicative Action Vol. 1: Reason and the Rationalisation of Society* (Cambridge: Polity Press in association with Basil Blackwell, Oxford, 1984).
23. Habermas, *Justification and Application*, 48–52.
24. Habermas, 'A Reply', 215.
25. ibid., 217.
26. Taylor, 'Language and Society', 23–36.
27. See Taylor, 'Ethics and Ontology', 305–310. In this article Taylor explores John McDowell's work with whom he concurs in the supposition that there is a need to move beyond the restrictive phenomenology of 'bald naturalism'. Taylor states, 'McDowell seems to have done the trick, and to have reconciled the deliverances of phenomenology (we really discern ethical differences in human life, and these have to be understood as involving incommensurable, higher values), and the basic concerns of a naturalistic ontology, which cannot allow such values into the furniture of the universe' (315).
28. Charles Taylor, 'A Catholic Modernity'. In *A Catholic Modernity?: Charles Taylor's Marianist Award Lecture*, edited by J. L. Heft (Oxford: Oxford University Press, 1999a) and Charles Taylor, 'Comment on Jürgen Habermas' 'From Kant to Hegel and Back Again', *European Journal of Philosophy*, 7(2) (1999): 158–160.
29. See for example J. Rawls, *Political Liberalism* (New York: Columbia University Press, 1993), and for Taylor's critique of Rawls' lexical approach to reconciliation see Charles Taylor, *Sources of the Self* (Cambridge: Cambridge University Press, 1989): 45.
30. Taylor, *Sources*, 45.
31. Habermas, 'A Reply', 217.
32. Immanuel Kant, *Groundwork on the Metaphysics of Morals* (Originally *The Moral Law*), translated by H. J. Paton (London: Hutchinson (now Harper and Row), 1949).
33. Jürgen Habermas, 'Kant's Idea of Perpetual Peace with the Benefit of Two Hundred Years' Hindsight'. In *Perpetual Peace: Essays on Kant's Cosmopolitan Ideal*, edited by J. Bohman and M. L. Lutz-Bachmann (London: MIT Press, 1997): 113–155.
34. Taylor, C., 'Comment on Jurgen Habermas' 'From Kant to Hegel and Back Again', *European Journal of Philosophy*, 7(2) (1999b): 158–160.
35. Charles Taylor, 'Foundationalism and the Inner-Outer Distinction'. In *Reading McDowell: On Mind and World*, edited by N. H. Smith (London and New York: Routledge, 2002): 106–121.
36. Terry Pinkard, *German Philosophy 1760–1860: The Legacy of Idealism* (Cambridge: Cambridge University Press, 2002), 227.

37. See especially Jürgen Habermas, *Moral Consciousness and Communicative Action*, Massachusetts Institute of Technology (Cambridge: Polity Press, 1990): 211.
38. Habermas, 'A Reply', 216.
39. Habermas, *Moral Consciousness and Communicative Action*, 211.
40. See Porter, *The Elusive Quest*, 154–155.
41. H. G. Gadamer, *Truth and Method*, Fifth reproduction, translated by Garrett Barden and John Cummings (London: Sheed and Ward, 1975).
42. Taylor, 'Engaged Agency and Background to Heidegger'. In *The Cambridge Companion to Heidegger*, edited by C. Guignon (Cambridge: Cambridge University Press, 1993): 317–337.
43. Johann G. Herder, 'On the Origin of Language'. In *J. G. Herder on Social and Political Culture*, translated and edited by F. M. Barnard (Cambridge: Cambridge University Press, 1969) and W. Von Humboldt, *On Language: The Diversity of Human Language-Structure and its Influence on the Mental Development of Mankind*, translated by P. Heath (Cambridge: Cambridge University Press, 1988). Taylor, 'The Importance of Herder', 79–100.
44. Taylor, 'Language and Society', 31.
45. Jürgen Habermas, *Between Facts and Norms* (Cambridge, MA: MIT Press, 1996): 108–109.
46. Taylor, *Sources*, 85–88. See also J. H. Zammito, *Kant, Herder, and The Birth of Anthropology* (Chicago: University of Chicago Press, 2002).
47. Charles Taylor, 'To Follow a Rule...'. In *Bourdieu: Critical Perspectives*, edited by C. Calhoun, E. LiPuma and M. Postone (Cambridge: Polity Press, 1993): 35–45 and Taylor, 'Engaged Agency and Background to Heidegger', 317–337.
48. Charles Taylor, *Hegel* (Cambridge: Cambridge University Press, 1975).
49. Taylor, *Sources ...*, 45.
50. Charles Taylor, 'The Dialogical Self'. In *The Interpretive Turn*, edited by D. R. Hiley, J. F. Bohman and R. Shusterman (Ithaca, NY: Cornell Press, 1991): 304–315.
51. Taylor, 'On Disclosing New Worlds', 119. See also M. N. Forster, 'Herder's Philosophy of Language, Interpretation, and Translation: Three Fundamental Principles', *The Review of Metaphysics*, LVI (2): (2002): 323–354.
52. Habermas, 'A Reply', 216–217.
53. Jürgen Habermas, 'Dialogue'. In *The Power of Religion in the Public Sphere*, edited by Judith Butler, Jürgen Habermas, Charles Taylor and Cornell West (New York: Columbia University Press, 2011): 61.
54. ibid.
55. ibid.
56. Annick Hedlund-deWitt, 'Worldviews and Their Significance for the Global Sustainability-Development Debate', *Environmental Ethics*, 35(2) (2013): 133–163.
57. See Taylor, 'Language and Society', 23–36. Also, Christina Lafont, *Heidegger, Language and World Disclosure* (Cambridge: Cambridge University Press, 2001): 11, 35.
58. Taylor, *The Ethics of Authenticity*, 10.
59. For Taylor's comments on Habermas's procedural *modus vivendi* see Taylor, 'Ethics and Ontology', 305.

60. Charles Taylor, 'Civil Society in the Western Tradition'. In *The Notion of Tolerance and Human Rights*, edited by E. Groffier and M. Paradis (Ottawa: Carleton University Press, 1991).

61. Charles Taylor, 'Alternative Futures: Legitimacy, Identity and Alienation in Late-Twentieth Century Canada'. In *Constitutionalism, Citizenship and Society in Canada*, edited by A. Cairns and C. Williams (Toronto: University of Toronto Press, 1985): 183–229 (reprinted in C. Taylor, 'Alternative Futures'. In *Reconciling the Solitudes*, edited by G. Laforest (Montreal: McGill, Queen's University Press 1993): 59–120.

62. Charles Taylor, 'A Tension in Modern Democracy'. In *Democracy and Vision*, edited by A. Botwinick and W. E. Connolly (Princeton and Oxford: Princeton University Press, 2001): 79–99.

63. Joel Anderson, 'The Personal Lives of Strong Evaluators: Identity, Pluralism, and Ontology in Charles Taylor's Value Theory', *Constellations* 3(1) (1996): 17–38, 31.

64. M. Ancelovici and F. Dupuis-Déri, 'Interview with Professor Charles Taylor', *Citizenship Studies*, 2(2) (1998): 251.

65. Charles Taylor, 'Understanding the Other: A Gadamerian View on Conceptual Schemes'. In *Gadamer's Century: Essays in Honor of Hans-Georg Gadamer*, edited by Jeff Malpas, U. Arnswald and Jens Kertshcer (Cambridge, MA: MIT Press, 2002): 290.

66. Michael Walzer, 'Comment', 'Struggles for Recognition in the Constitutional Democratic State'. In *Multiculturalism*, edited by Amy Gutmann (Princeton: Princeton University Press, 1994): 99.

67. Ruth Abbey, 'Interpreting Taylor Rightly', *Ethnicities*, 31(1) (2003): 123.

68. Michael Mesner, 'What Might Environmental Philosophy Learn from Iris Young's Metropolitan Regionalism', *Environmental Ethics*, 35(4) (2013): 439–461.

69. M. Patrick, 'Liberalism, Rights and Recognition', *Philosophy and Social Criticism*, 26(5) (2000): 28–4 37.

70. C. Taylor, 'Dialektik heute, oder: Strukturen der Selbstnegation', In *Hegel's Wissenchaft der Logik: Formation und Rekonstruktion*, edited by D. Henrich (Stuttgart: Klett-Cotta, 1986): 141–153.

71. ibid.

72. ibid., 141.

73. ibid.

74. Taylor, *Hegel*, 550.

75. Taylor, 'Understanding the Other', 290.

76. Charles Taylor, 'Explanation and Practical Reason'. In *The Quality of Life*, edited by Martha Nussbaum and Amartya Sen (Oxford: Clarendon Press, 1983): 208–209.

77. Charles Taylor, 'Neutrality in Political Science'. In *The Philosophy of Social Explanation*, edited by A. Ryan (Oxford: Oxford University Press, 1973): 139–171.

78. Russell Hittinger, 'Charles Taylor: *Sources of the Self*', *Review of Metaphysics*, 44(1) (1990): 111–130.

79. ibid.

80. Redhead, *Charles Taylor: Thinking and Living Deep Diversity* (Maryland: Rowman & Littlefield, 2002), 203.

81. Mark Redhead, 'Charles Taylor's Nietzschean Predicament: A Dilemma More Revealing Than Foreboding', *Philosophy and Social Criticism*, 27(6) (2001): 100.
82. Hittinger, 'Charles Taylor: *Sources of the Self*', 111–130.
83. Bernard Williams, 'Republican and Galilean', *New York Review of Books* (8 November, 1990): 48.
84. Redhead, 'Charles Taylor's Nietzschean Predicament: A Dilemma More Revealing Than Foreboding', 98.
85. ibid.
86. Ruth Abbey, *Nietzsche's Middle Period* (Oxford: Oxford University Press, 2000).
87. ibid., 18.
88. Taylor, 'Understanding the Other', 279–299.
89. Gadamer, *Truth and Method*, 306–307, 374–375.
90. Stephen K. White, *Sustaining Affirmation* (Princeton: Princeton University Press, 2000): 68.
91. Stephen K. White, 'Fullness and Dearth: Depth Experience and Democratic Life', *American Political Science Review*, 104(4) (2010): 801.
92. ibid.
93. ibid.
94. Taylor, 'Heidegger, Language, and Ecology', 247–269.
95. ibid., 257.
96. In this sense, Taylor's interpretation extends Gadamer's argument that it is through language that people's moral sources can also be given expression to satisfy people's need for recognition. For environmental purposes, as an example, it explores that ground where nature's value can be revealed as it shapes the development of freedom. See Taylor, 'Understanding the Other', 290.
97. Taylor, 'A Tension in Modern Democracy', 79–99.
98. Patrick, 'Liberalism, Rights and Recognition', 28–47.
99. Taylor, 'Cross-Purposes: The Liberal-Communitarian Debate', 159–182.
100. Charles Taylor, *Multiculturalism and the Politics of Recognition*, edited by Amy Gutmann (Princeton, NJ: Princeton University Press, 1992): 25; (expanded edition of *Multiculturalism: Examining the Politics of Recognition*, 1992).
101. John S. Dryzek, 'Democratization as Deliberative Capacity Building', *Comparative Political Studies*, 42 (November 2009), 1381–1382.
102. Robyn Eckersley, *The Green State* (Cambridge, MA: MIT Press, 2004); Robyn Eckersley, *Globalization and the Environment* (London: Rowman & Littlefield, 2013).
103. ibid., 91–92.
104. Eckersley, *The Green State*, 111.
105. Robyn Eckersley, 'Climate Challenge for Rudd', *The Age*, 3 December 2007.
106. ibid.
107. Eckersley, *The Green State*, 166.
108. Robyn Eckersley, *Green Theory. International Relations Theories: Discipline and Diversity* (Oxford: Oxford University Press, 2013): 266–286.
109. Hubert Dreyfus, 'Responses'. In *Heidegger, Coping and Cognitive Science: Essays in Honor of Hubert L. Dreyfus,* edited by M. A. Wrathall and J. Malpas (Cambridge, MA: MIT Press, 2000): 344.

9 Critical Perspectives: The Taylor–Rorty Debate

1. Often instrumental theories are aligned with a version of utilitarianism that reflects an economic approach to the natural environment. As a consequence, the natural environment has been treated as a commodity and its non-instrumental values submerged by a utilitarian social imaginary.
2. Charles Taylor, 'Politics of the Steady State', *New Universities Quarterly* 32(2) (1978): 157–184.
3. Johnathan Marks, 'Misreading One's Sources: Charles Taylor's Rousseau', *American Journal of Political Science*, 49(1): (2005): 119–134.
4. Charles Taylor, 'The Significance of Significance: The Case of Cognitive Psychology'. In *The Need for Interpretation: Contemporary Conceptions of the Philosopher's Task*, edited by S. Mitchell and M. Rosen (London: Athlone Press, 1983): 141–169.
5. Marks, 'Misreading One's Sources', 130.
6. Jürgen Habermas, 'Struggles for Recognition in the Constitutional Democratic State'. In *Multiculturalism*, translated by Shierry Weber Nicholsen, edited by Amy Gutman (Princeton: Princeton University Press, 1994), 130–131.
7. Taylor, *Sources*, ch. 20.
8. Charles Taylor, 'Comment on Jürgen Habermas' 'From Kant to Hegel and Back Again', *European Journal of Philosophy*, 7(2) (1999): 161.
9. Marks, 'Misreading One's Sources', 130.
10. Mark Wolin, *Heidegger's Children* (Princeton and Oxford: Princeton University Press, 2001) and T. Rockmore, *On Heidegger's Nazism and Philosophy* (Berkeley, CA: University of California Press, 1997).
11. Ruth Abbey, 'More Perspectives on Communitarianism', *Australian Quarterly*, 69(2): (1997): 1–11; R. Abbey, *Charles Taylor* (Oxford: Oxford University Press, 2001), and Ruth Abbey, 'Taylor as a Postliberal Theorist of Politics'. In *Perspectives on the Philosophy of Charles Taylor*, edited by A. Laitinen and N. Smith (*Acta Philosophica Fennica* 71, Helsinki: The Philosophical Society of Finland 2002): *passim*.
12. Charles Taylor, 'Response to my Commentators', *Philosophy and Phenomenological Research*, 54(1) (1994): 213.
13. Charles Taylor, *A Secular Age* (Cambridge, MA: Harvard University Press, 2007): ix.
14. Interpretivists differ from environmentalists such as J. B. Callicott (see J. B. Callicott and J. Baird, *Earth's Insights: A Multicultural Survey of Ecological Ethics from the Mediterranean Basin to the Australian Outback* (Berkeley and Los Angeles: The Regents of the University of California, 1994) who refers to himself as a communitarian (environmentalist). However, the natural environment is more than an entity with intrinsic value – it is a necessary condition of our being in the world.
15. Jean Baudrillard, *The Illusions of the End* (Stanford: Stanford University Press, 1994), 82.
16. Taylor, *Sources*, 358. Here Taylor is quoting Rousseau, *Emile*, 355 (Rousseau, Jean-Jacques, *Emile*, translated by Allan Bloom (New York: Basic Books, 1979).
17. Charles Taylor, 'Statement', http://www.templetonprize.org/ct_statement.html, accessed 21 January 2012.

18. Richard Rorty, 'Robert Brandom on Social Practices'. In *Truth and Progress: Philosophical Papers*, Vol. 3, edited by Richard Rorty (Cambridge: Cambridge University Press, 1998): 84–97, 127.
19. Taylor, *Sources*, 358.
20. Heidegger famously stated that it is language that speaks. I argue that Heidegger's allegedly notorious phrase can be interpreted to mean that interpretation and understanding are an integral feature of language.
21. H. Dreyfus, 'Taylor's (Anti-) Epistemology'. In *Charles Taylor*, edited by R. Abbey (Cambridge: Cambridge University Press, 2004): 55.
22. Richard Rorty, *Objectivity, Relativism and Truth: Philosophical Papers*, Vol. 2 (Cambridge: Cambridge University Press, 1991): 6.
23. Richard Rorty, *Philosophy and the Mirror of Nature* (Princeton, NJ: Princeton University Press, New Jersey, 1979): 176.
24. ibid., 174.
25. ibid.
26. Richard Rorty, 'Charles Taylor on Truth'. In *Truth and Progress: Philosophical Papers Volume 3*, edited by Richard Rorty (Cambridge: Cambridge University Press, 1998), 93.
27. ibid.
28. Dreyfus, 'Taylor's (Anti-) Epistemology', 75.
29. Taylor, 'Comment on Jurgen Habermas' 'From Kant to Hegel and Back Again'', 161.
30. Hubert Dreyfus, 'Holism and Hermeneutics', *Review of Metaphysics*, 34(1) (1980): 19.
31. Charles Taylor, 'Understanding in the Human Sciences', *Review of Metaphysics*, 34(1) (1980): 30.
32. Heidegger, Martin, 'The Thing'. In Poetry, *Language, Thought, A Hofstadter* (Trans). (New York, Harper Rowe, 1971): 170. Emphasis in original. Found originally in Dreyfus, 'Taylor's (Anti-) Epistemology, 78.
33. Dreyfus, 'Holism and Hermeneutics', 16–17. More recently, Hubert Dreyfus and Charles Spinosa, 'Coping with Things-in-Themselves: A Practice-Based Phenomenological Argument for Research', *Inquiry*, 42(2) (1999): 49–78. See also Taylor, 'Understanding in the Human Sciences', 25–38.
34. H. L. Dreyfus ('Taylor's (Anti-) Epistemology', In *Charles Taylor*, edited by R. Abbey (Cambridge: Cambridge University Press, 2004), 60–63) notes that Taylor's most recent work can answer the brain in the vat objection. In a possible world it might be the case that we are not really embodied or coping agents engaged with the world because we are given the impression that we are outside agents. No matter how unlikely this scenario is, Dreyfus notes that it can be met by noting that it is an agent's perception that matters. It seems that Rorty glides over Taylor's work on perception and knowledge.
35. Heidegger, M., *Poetry, Language, Thought* (New York, NY: Harper & Rowe, 1971).
36. Heidegger, 'The Thing' in *Poetry, Language, Thought*, 170.
37. Taylor in Dreyfus, 'Taylor's (Anti-) Epistemology', 75.
38. John McDowell, *Mind and World* (Cambridge, MA: Harvard University Press, 1993): 58.
39. Taylor in Dreyfus, 'Taylor's (Anti-) Epistemology', 58.
40. See Rorty, Charles Taylor on Truth', 96.
41. Gadamer, *Truth and Method*, 378–379.

42. Charles Taylor, *Multiculturalism and the Politics of Recognition*, edited by Amy Gutmann (Princeton, NJ: Princeton University Press, 1992): 25; (expanded edition of *Multiculturalism: Examining the Politics of Recognition*, 1992); Charles Taylor, *Varieties of Religion Today* (Harvard: Harvard University Press, 2002); Charles Taylor, 'Understanding the Other: A Gadamerian View on Conceptual Schemes'. In *Gadamer's Century: Essays in Honor of Hans-Georg Gadamer*, edited by Jeff Malpas, U. Arnswald and Jens Kertshcer (Cambridge, MA: MIT Press, 2002): 290.

43. Alison Ross, *The Aesthetic Paths of Philosophy: Presentation in Kant, Heidegger, Lacou-Labeth, and Nancy* (Stanford: Stanford University Press, 2007): 96.

44. As quoted in E. Palmer, *Hermeneutics, Interpretation Theory in Schleiermacher, Dilthey, Heidegger and Gadamer* (Evanston: Northwestern University Press, 1969): 137–8 (Palmer here cites Martin Heidegger, *Being and Time* (Cambridge, MA: MIT Press, 1994): 157–8).

45. Charles Taylor, 'Heidegger, Language and Ecology'. In *Philosophical Arguments*, edited by Charles Taylor (Cambridge, MA: Harvard University Press, 1995): 263.

46. G. Gadamer, *Truth and Method*, Fifth reproduction, translated by Garrett Barden and John Cummings (London: Sheed and Ward, 1975): 50–51.

47. Lars Samuelson, 'On the Possibility of Evidence for Intrinsic Value in Nature', *Ethics and the Environment*, 18(2) (2013): 111.

48. ibid.

49. ibid.

50. Rorty, 'Charles Taylor on Truth', 93–94.

51. Taylor offers the example concerning how we engage with things in the world. He explores how we check whether the mirror is crooked or straight. Taylor explains that of course, we check our claims against reality. And we do so without relying solely on our conscious representations. 'Johnny go into the room and tell me whether the picture is crooked'. Johnny emerges from the room with a view of the matter, but checking is not comparing the problematised belief with his belief about the matter; checking is forming a belief about the matter, in this case by going and looking. What is assumed when we give the order is that Johnny knows, as most of us do, how to form a reliable view of this kind of matter. He knows how to go and stand at the appropriate distance and in the right orientation, to get what Merleau-Ponty calls a maximal grip on the object. What justifies Johnny's belief is his being able to deal with objects in this way, which is, of course, insepa-rable from the other ways he is able to use them, manipulate, get around among them, etc. When he goes and checks he uses this multiple ability to cope, and his sense of his ability to cope gives him confidence in his judgement as he reports it to us (Taylor, quoted in Dreyfus, 'Taylor's (Anti-) Epistemology', 56

52. Taylor defined supersession arguments as framed by ontological parameters; this is the process of moving from confusion to clarity through a process that begins by recognising the different ways different people visualise the world.

53. Dreyfus, 'Taylor's (Anti-) Epistemology', 69.

54. C. Taylor, 'Reply and Rearticulation: Charles Taylor Replies'. In *Philosophy in an Age of Pluralism: The Philosophy of Charles Taylor in Question*, edited by C. Tully (Cambridge, Cambridge University Press, 1994), 222.

55. Taylor, 'Understanding in the Human Sciences', 27.
56. Taylor, 'Heidegger, Language and Ecology', 262.
57. Heidegger, *'The Thing'* in *Poetry, Language, Thought';* Martin Heidegger, 'Building Dwelling Thinking' from *Poetry, Language, Thought*, translated by Albert Hofstadter (New York: Harper Colophon Books, 1971): 141–161. Found in Dreyfus, 'Taylor's (Anti-) Epistemology', 77.
58. ibid., 79.
59. D. A. Brown, 'The Role of Law in Sustainable Development and Environmental Protection Decision Making'. In *Sustainable Development: Science, Ethics and Public Policy*, edited by J. Lemons and D. A. Brown (Dordrecht: Kluwer Academic Press, 1995): 67.
60. Taylor, *A Secular Age*, part I.

10 Taylor and Deep Ecology

1. David Henry Thoreau, *Walking* (1862). Originally published in *The Atlantic Monthly*, 9(56): 657–674. Copy found at http://www.wildnesswithin. com/2001/01-12/inwildness.html (accessed 9 March 2014).
2. Charles Taylor, 'Language Not Mysterious?' In *Reading Brandom: On Making It Explicit*, edited by Bernhard Weiss and Jeremy Wanderer (London: Routledge, 2010): 32–47.
3. See Warwick Fox, 'The Deep Ecology-Ecofeminism Debate and Its Parallels', *Environmental Ethics*, 11(1) (1980): 5–26; W. Fox, 'Deep Ecology: A New Philosophy for Our Time', *The Ecologist*, 14(5–6) (1984): 194–200 and W. Fox, *Toward a Transpersonal Ecology* (Boston, MA: Shambala, 1990). For feminist perspectives see A. Salleh, 'Deeper Than Deep Ecology: The Eco-Feminist Connection', *Environmental Ethics*, 6(4) (1984): 339–345. For feminist perspectives see A. Salleh, 'The Ecofeminism/Deep Ecology Debate', *Environmental Ethics*, 14(3) (1992): 195–216; A. Salleh, 'Class, Race, and Gender Discourse in the Ecofeminism/Deep Ecology Debate', *Environmental Ethics*, 15(3) (1993): 225–245.
4. Bron Taylor, 'Earth and Nature-Based Spirituality (Part 1): From Deep Ecology to Radical Environmentalism', *Religion*, 31(2) (2001): 180.
5. ibid., 175.
6. Bill Devall and George Sessions, *Deep Ecology* (Salt Lake City, UT: Peregrine Smith Books, 1985): 75.
7. Deep ecologists argue the following: (1) The well-being and flourishing of human and nonhuman Life on Earth have value in themselves (synonyms: intrinsic value, inherent value). These values are independent of the usefulness of the nonhuman world for human purposes (2) Richness and diversity of life forms contribute to the realisation of these values and are also values in of themselves (3) Humans have no right to reduce this richness and diversity except to satisfy vital needs (4) The flourishing of human life and cultures is compatible with a substantial decrease of the human population. The flourishing of nonhuman life requires such a decrease (5) Present human interference with the nonhuman world is excessive, and the situation is rapidly worsening (6) Policies must therefore be changed. These policies affect basic economic, technological, and ideological structures.

The resulting state of affairs will be deeply different from the present (7) The ideological change is mainly that of appreciating life quality (dwelling in situations of inherent worth) rather than adhering to an increasingly higher standard of living. There will be a profound awareness of the difference between big and great (8) Those who subscribe to the foregoing points have an obligation directly or indirectly to participate in the attempt to implement the necessary changes.

8. Tom Regan, *The Case for Animal Rights* (London: Routledge & Kegan, 1983): 361–362.

9. Brian Barry, 'Review: Sandel, Michael J. *Liberalism and the Limits of Justice*', *Ethics*, 94(3) (1984): 523–524.

10. Brian Barry's moral and political philosophy has made the point that Michael Sandel's work in the field of interpretation was likely to lead to a reign of terror on citizens. Barry states that Sandel makes the transcendence of justice by group identity sound very high-minded. However, it gives the green light to every string-pulling parent and crony-hiring academic. At the end of that road, Torquemada, Stalin, Hitler and Begin stand. Sandel's argument should be turned on its head: it is exactly when 'devotion to city or nation, to party or cause' run deepest that the constraints of justice on the pursuit of those allegiances are most needed (B. Barry, 'Review: Sandel, Michael J. *Liberalism and the Limits of Justice*', *Ethics*, 94(3) (1984): 523–524.

11. Charles Taylor, *Hegel* (Cambridge: Cambridge University Press, 1975): 547.

12. Charles Taylor, 'Heidegger, Language, and Ecology'. In *Heidegger: A Critical Reader*, edited by H. L. Dreyfus and H. Hall (Oxford: Blackwells, 1992): 247–269.

13. Michael E. Zimmerman, 'Rethinking the Heidegger – Deep Ecology. Relationship', *Environmental Ethics*, 15(3) (1993): 195–224.

14. Jacques Derrida, '*Différance*'. In *Margins of Philosophy*, translated by Alan Bass (Chicago: University of Chicago Press, 1982). That is, Heidegger was operating with a particular form of species-based politics. Jacques Derrida, a postmodern philosopher whose main focus was on binary oppositions and deconstruction, suggested that even if Heidegger 'was not guilty of biological racism, perhaps he was guilty of a type of "metaphysical" racism.' Jacques Derrida noted that the key problem with Heidegger's politics is the way in which they emphasised German linguistic superiority. Critics continue to apply this criticism to Deep Ecology.

15. According to Taylor's 'Heidegger, Language, and Ecology', 100, the notion of Dasein is a way of involvement and caring for the immediate world while remaining aware of the contingent element of that involvement as well as the priority of the world to the self, and of the evolving nature of the self itself.

16. Charles Taylor, 'Ethics and Ontology', *Journal of Philosophy*, C(6) (2003): 307.

17. Daniel Baird, 'Charles Taylor's *A Secular Age*, 12 September 2012, n.p.

18. ibid.

19. ibid.

20. ibid.

21. L. P. Hinchman, 'Aldo Leopold's Hermeneutic of Nature', *The Review of Politics*, 57(2) (1995): 247.

22. Jim Thompson, 'A Refutation of Environmental Ethics', *Environmental Ethics*, 12(2) (1990): 147–160.

23. See Laura Westra, 'Let It Be: Heidegger and Future Generations', *Environmental Ethics*, 7(4) (1985): 341–350, and Laura Westra, 'Ecology and Animals: Is There a Joint Ethic of Respect', *Environmental Ethics*, 11(3) (1989): 215–260.

11 Taylor's Evaluative Framework and Critical Perspectives

1. For critical perspectives on Taylor see Charles Thiebaut, 'Charles Taylor: On the Improvement of our Moral Portrait (Moral Realism, History of Subjectivity and Expressivist Language)', *Praxis International*, 13(2) (1993): 126–154.
2. Jürgen Habermas, 'Struggles for Recognition in the Constitutional Democratic State'. In *Multiculturalism*, translated by Shierry Weber Nicholsen, edited by Amy Gutman (Princeton: Princeton University Press, 1994), 130–131.
3. Wendy Brown, 'The Sacred, The Secular, and The Profane: Charles Taylor and Karl Marx'. In *Varieties of Secularism in a Secular Age*, edited by Michael Warner, Jonathan Van Antwerpen and Craig Calhoun (Harvard: Harvard University Press, 2010).
4. For a critical overview of Frazer's work see Ruth Abbey, 'Turning or Spinning? Charles Taylor's Catholicism', *Contemporary Political Theory* 5(2) (2006): 163–175.
5. Anderson, 'The Personal Lives of Strong Evaluators: Identity, Pluralism, and Ontology in Charles Taylor's Value Theory', *Constellations: An International Journal of Critical and Democratic Theory*, 3 (1996): 17–38, 33; Smith, 'Review Essay: Reason after Meaning: Charles Taylor, Philosophical Arguments', *Philosophy and Social Criticism*, 23(1): 131–140; Vicki Spencer, 'Towards an Ontology of Holistic Individualism: Herder's Theory of Identity, Culture and Community', *History of European Ideas*, 22(3) (1996): 245–260.
6. Y. Huang, 'Charles Taylor's Transcendental Arguments for Liberal Communitarianism', *Philosophy and Social Criticism*, 24(4) (1998): 79–107; see Hermut Rosa, 'Cultural Relativism and Social Criticism from a Taylorian Perspective', *Constellations*, 3(1) (1996): 39–57; Judith Shklar, 'Review of Sources of the Self', *Political Theory*, 19(1) (1991): 105–109; Bernard Williams, 'Republicanism and Galilean', *The New York Review of Books* (8 November 1990): 45–48.
7. See Nicholson, 'To Be or Not To Be: Charles Taylor and the Politics of Recognition', *Constellations*, 3 (April 1996): 1–16.
8. See for example Soper's critique of D. J. Haraway, *Simians, Cyborgs, and Women: The Reinvention of Nature* (London: Free Association Books, 1991). Soper, K., 'The Limits of Hauntology', *Radical Philosophy*, 75 (1996): 30 and K. Soper, 'The Goodness of Nature and the Nature of Goodness', *Capitalism, Nature and Socialism*, 11(1) (2000): 87–93.
9. Isaiah Berlin, 'Introduction'. In *Philosophy in an Age of Pluralism: The Philosophy of Charles Taylor in Question*, edited by J. Tully (Cambridge: Cambridge University Press, 1994): 1–3.
10. ibid., 2.
11. Charles Taylor, 'Reply and Rearticulation: Charles Taylor Replies'. In *Philosophy in an Age of Pluralism: The Philosophy of Charles Taylor in Question*, edited by J. Tully. (Cambridge: Cambridge University, 1994), Isaiah Berlin,

'Introduction'. In *Philosophy in an Age of Pluralism: The Philosophy of Charles Taylor in Question*, edited by J. Tully (Cambridge: Cambridge University Press, 1994): 213–258.

12. Charles Taylor, *Hegel* (Cambridge: Cambridge University Press, 1975): 550.
13. For a detailed analysis of Marx and the environment see J. Hughes, *Ecology and Historical Materialism* (Cambridge: Cambridge University Press, 2000).
14. Karl Marx in *Karl Marx: Selected writings*, edited by David McLellan (Oxford: Oxford University Press, 1977.
15. Taylor, *Hegel*, 555.
16. ibid.
17. Taylor's analysis of capitalism agrees with that of B. Fontana, 'The Concept of Nature in Gramsci', *The Philosophical Forum*, 27(3) (1996): 220–244.
18. For Taylor's interpretations of Marx see Charles Taylor, 'Socialism and Weltanschauung'. In *The Socialist Idea: A Reappraisal*, edited by L. Kolakowski and S. Hampshire (New York: Basic Books, 1974); C. Taylor, 'Marxism and Empiricism'. In *British Analytic Philosophy*, edited by B. Williams and A. Montefiore (London: Routledge and Kegan Paul, 1966); Charles Taylor, 'Two Issues about Materialism', *The Philosophical Quarterly*, 19(74) (1969): 73–79; Charles Taylor, 'The Logics of Disintegration', *The New Left Review*, 170 (1988): 110–118.
19. Charles Taylor, 'What Is Human Agency'. In *Philosophical Papers*, Vol. 1, edited by Charles Taylor (Cambridge: Cambridge University Press, 1985): 25–26.
20. Taylor, *Sources*, 63.
21. See the Introduction and Chapter 6 for an analysis of strong evaluations and hyper-goods.
22. C. Taylor, 'Dialektik heute, oder: Strukturen der Selbstnegation', In *Hegel's Wissenchaft der Logik: Formation und Rekonstruktion,*edited by D. Henrich (Stuttgart: Klett-Cotta, 1986): 141.
23. ibid.
24. ibid.
25. ibid.
26. Taylor, 'The Logics of Disintegration', 10–118.
27. Taylor, *Hegel*, 553.
28. ibid., 550–555.
29. A. Swingewood, *Marx and Modern Social Theory* (London: Macmillan Press, 1975).
30. Wendy Brown, 'The Sacred, The Secular, and The Profane',
31. Taylor, *Hegel*, 553.
32. Adapted from Tony Smith, *The Logic of Marx's 'Capital': Replies to Hegelian Criticisms* (Albany: State University of New York Press, 1993).
33. B. Brugger and D. Kelly, *Chinese Marxism in the Post-Mao Era* (California: Stanford University Press, 1990), 62.
34. Karl Marx, *Grundrisse: Foundations of the Critique of Political Economy (Rough Draft),*(Harmondsworth: Penguin, 1973), 100.
35. G. W. F. Hegel, *Hegel's Logic,* translated by William Findlay (Oxford: Oxford University Press, 1975): 116.
36. G. W. F. Hegel, *The Science of Logic*, edited by A.V. Miller (London: George Allen and Unwin, 1969): 828–829.
37. Karl Marx, *Letters on 'Capital'* (London: New Park, 1893): 129.

38. His analysis began with the concepts of *species being* and *abstract labour* taken together, whereby the former offered an anthropological understanding of human existence while the latter expressed labour-power in the commodity form. It was through the categories of abstract labour and species being that Marx captured the essence of capitalism rather than constructing a negative critique.

39. Karl Marx and Friedrich Engels, *Collected Works*, Vol. 3 (Moscow: Progress Publishers, 1975): 32.

40. Marx, *Letters on 'Capital'* (London: New Park, 1893): 148. Initially found in Tony Smith, *The Logic of Marx's 'Capital': Replies to Hegelian Criticisms* (Albany: State University of New York Press, 1993): 70.

41. K. Marx, 'The Critique of the Gotha Programme'. In *Selected Works in Two Vols*, Vol. II, edited by K. Marx and F. Engels (Moscow: Foreign Languages Publishing House, 1951): 13–35.

42. R. Grundmann, *Marxism and Ecology* (Oxford: Clarendon Press, 1991) and R. Grundmann, 'The Ecological Challenge to Marxism', *New Left Review*, 187 (May–June 1991): 103–120.

43. K. Marx, 'Economic and Philosophical Manuscripts: Second Manuscript – Private Property and Communism, *Early Writings*, edited by T. B. Bottomore (London and New York: C. A Watts and Company, 1964): 155. Initially found in Taylor, *Hegel*, 550.

44. K. Marx in Marx-Engels, *Collected Works*, Vol. 10 (London: Lawrence and Wishart, 1976): 245. Initially found in R. Grundmann, 'The Ecological Challenge to Marxism', *New Left Review*, 187 (May–June 1991): 110.

45. A. Light, 'Rereading Bookchin and Marcuse as Environmental Materialists', *Capitalism, Socialism and Nature*, 4(1) (1993): 69–98.

46. Of course, in other parts of his work, such as his *1844 Manuscripts*, he explains the importance of the laws of nature in understanding beauty and value. He states: 'An animal forms objects only in accordance with the standard and the need of the species to which it belongs, whilst man knows how to produce in accordance with the standard of every species, and knows how to apply everywhere the *inherent standard* to the object. Man therefore also forms objects in accordance with the laws of beauty' (Marx and Engels, *Collected Works*, 32).

47. C. Taylor, 'Review: The Primacy of Perception: And Other Essay on Phenomenological Psychology, the Philosophy of Art, and History' by Maurice Merleau-Ponty, edited, with an introduction by James M. Edie (Evanston: Northwestern University Press, 1964): xix, 228. By Maurice Merleau-Ponty, translated, with an Introduction, Richard C. McCleary (Evanston: Northwestern University Press, 1964): xxxiv, 355. Maurice Merleau-Ponty's work Source: *The Philosophical Review*, 76(1) (January 1967): 117

48. Charles Taylor interviewed by Ben Rogers, *Prospect Magazine* (29 February 2008), Issue 143, 15.

49. Carol Pateman, *The Sexual Contract* (Oxford: Basil Blackwell, 1988). She argues that participation is not simply a matter of better political participation, but also better participation in labour relations.

50. For a critique of consumerism, see Jean Baudrillard, *For a Critique of the Political Economy of the Sign* (St. Louis: Telos, 1981); Ivan Illich, *Energy and*

Equity (New York: Harper, 1974); William Leiss, *The Limits of Satisfaction* (Toronto: University of Toronto Press, 1976); and Timothy W. Luke, 'Regulating the Haven in a Heartless World: The State and Family under Advanced Capitalism', *New Political Science*, 2(3) (1981): 51–74.

51. Luke, *Capitalism, Democracy and Ecology*, 197–198.
52. Charles Taylor, 'What's Wrong with Foundationalism?: Knowledge, Agency and World'. In *Heidegger, Coping and Cognitive Science*, edited by M. A. Wrathall and J. Malpas (Massachusetts: University of Massachusetts Press, 2000): 115–134.
53. Taylor, *Hegel*, 550.

12 Conclusion

1. Charles Taylor, *A Secular Age* (Cambridge, MA: Harvard University Press, 2007): 26.
2. Timothy Moyle, 'Re-Enchanting Nature: Human and Animal Life in Later Merleau-Ponty', *Journal of the British Society for Phenomenology*, 38(2) (2007): 178.
3. Charles Taylor, 'Explanation and Practical Reason'. In *The Quality of Life*, edited by Martha Nussbaum and Amartya Sen (Oxford: Clarendon Press, 1983): 209.
4. ibid.
5. H. Dreyfus, 'Taylor's (Anti-) Epistemology'. In *Charles Taylor*, edited by R. Abbey (Cambridge: Cambridge University Press, 2004): 68.

Index

CPSIA information can be obtained at www.ICGtesting.com
Printed in the USA
LVOW04*0259150815

450256LV00007B/176/P